HIDDEN IN PLAIN SIGHT

HIDDEN IN PLAIN SIGHT

The Social Structure of Irrelevance

EVIATAR ZERUBAVEL

OXFORD
UNIVERSITY PRESS

Oxford University Press is a department of the University of Oxford.
It furthers the University's objective of excellence in research, scholarship,
and education by publishing worldwide.

Oxford   New York
Auckland   Cape Town   Dar es Salaam   Hong Kong   Karachi
Kuala Lumpur   Madrid   Melbourne   Mexico City   Nairobi
New Delhi   Shanghai   Taipei   Toronto

With offices in
Argentina   Austria   Brazil   Chile   Czech Republic   France   Greece
Guatemala   Hungary   Italy   Japan   Poland   Portugal   Singapore
South Korea   Switzerland   Thailand   Turkey   Ukraine   Vietnam

Oxford is a registered trade mark of Oxford University Press
in the UK and certain other countries.

Published in the United States of America by
Oxford University Press
198 Madison Avenue, New York, NY 10016

© Eviatar Zerubavel 2015

All rights reserved. No part of this publication may be reproduced,
stored in a retrieval system, or transmitted, in any form or by any means,
without the prior permission in writing of Oxford University Press,
or as expressly permitted by law, by license, or under terms agreed with the
appropriate reproduction rights organization. Inquiries concerning reproduction
outside the scope of the above should be sent to the Rights Department,
Oxford University Press, at the address above.

You must not circulate this work in any other form
and you must impose this same condition on any acquirer

Library of Congress Cataloging-in-Publication Data
Zerubavel, Eviatar.
  Hidden in plain sight : the social structure of irrelevance / Eviatar Zerubavel.
       p. cm.
  Includes bibliographical references and index.
  ISBN 978-0-19-936660-6 (hardcover : alk. paper) — ISBN 978-0-19-936661-3
(pbk. : alk. paper) 1. Meaning (Psychology) 2. Relevance. 3. Attention. 4. Social
structure. I. Title.
  BF778.Z47 2015
  302'.12—dc23
                              2014022251

9 8 7 6 5 4 3 2 1

Printed in the United States of America on acid-free paper

To Peter Berger,
who opened for me the door to social phenomenology

And the eyes of them both were opened, and they knew that they were naked.

GENESIS 3:7

CONTENTS

	Preface	ix
1.	Noticing and Ignoring	1
2.	Figure and Background	11
	Relevance and Irrelevance	20
	The Marked and the Unmarked	22
3.	Searching and Hiding	24
	Spotting	24
	In the Background	27
	Background Matching	30
	Contour Distortion	37
	Diversion	44
4.	The Social Organization of Attention	49
	Nature and Culture	49
	Socio-Attentional Patterns	53
	Norms and Control	59
	Attentional Socialization	63
	Collective Attention	69
5.	Conclusion	72
	Multifocal Attention	75
	Open Awareness	77

Relevance Reconsidered 79
Foregrounding 82
Beyond Figure and Background 89

Notes *95*
Bibliography *151*
Author Index *185*
Subject Index *193*

PREFACE

My great fascination with the concepts "figure" and "background" dates back to 1969, when I first read about the Gestalt theory of perception in college. Seven years later, in my doctoral dissertation, I began to realize their potential extra-psychological use as I was analyzing the question "What are you doing here?" we often ask people we see at unexpected times,[1] and I continued to explore such odd figure-ground configurations in my 1981 book *Hidden Rhythms*.[2]

But the main theme of the present book dates back to a 1993 article[3] I later developed into a chapter on the social organization of attention in my 1997 book *Social Mindscapes*.[4] It was there that I first noticed the connection between the figure-and-ground model of perception and the social structure of relevance. Later, in my 2006 book on conspiracies of silence, *The Elephant in the Room*, I began to also appreciate the importance of "foregrounding" as a cognitively subversive way of combating denial.[5]

Then, on December 14, 2009, I happened to watch the Chamber Music Society of Lincoln Center playing Johann Sebastian Bach's Brandenburg concertos. As soon as the concert began, I was immediately drawn to the sounds produced by the double-bass player Edgar Meyer, and then realizing that I was actually attending to the part of the music that is conventionally supposed to remain "in the background"! It was my familiarity with the piece, of course, that allowed me to focus on the "accompaniment" while suspending my attention to the "melody." A few days later, while telling a friend about that experience, I decided to write a book about "the background."

Several people played a major role in my efforts to produce this book. I am particularly grateful to my family – Yael Zerubavel, Noga Zerubavel, Noam Zerubavel, and Dave Waller, as well as to my former students Tom DeGloma, Asia Friedman, and Ruth Simpson, who read my manuscript and spent many hours discussing it with me. I also benefited tremendously from some very useful feedback I got from several other friends, students, and colleagues who read earlier drafts of the manuscript – Caroline Levisse, Alicia Raia, Wayne Brekhus, Ira Cohen, Christena Nippert-Eng, Zali Gurevich, Ian Watson, Doug Harper, and Simon Gottschalk. I also wish to thank Ilanit Palmon and Eiko Saeki for their help with some of the illustrations in the book, and James Cook for his sound editorial advice.

I dedicate this book to Peter Berger, who forty-three years ago opened for me the door to the enchanting world of social phenomenology.

East Brunswick, New Jersey, 2014

HIDDEN IN PLAIN SIGHT

NOTICING AND IGNORING

> I remember taking one of the wilderness classes I teach out for a walk. We passed a dozen deer, two foxes, one cottontail, six groundhogs, a myriad of birds, insects and other creatures. Nobody noticed even one of them. When I went through the list, the students were angry at themselves: how could they have missed so much?
>
> TOM BROWN JR.[1]

As you are reading this, you too, indeed, are most likely unaware of the constant sound of your breathing, your own body odor, or the persistent pressure of the shoes you are wearing against your feet. Nor is it unusual to suddenly realize that something for which you have been searching for some time was in fact right in front of you the whole time, hidden, as they say, in plain sight.

Such *inattentiveness*, whereby things that can easily penetrate our consciousness nevertheless remain unperceived,[2] is in fact a very common phenomenon.[3] Indeed, what we can access through our sense organs and thus potentially see, hear, taste, or smell is always more than what we actually notice. Simply looking at something, for instance, does not guarantee, therefore, that you will in fact notice it.[4]

Such fundamental discrepancy between what is perceptually accessible to us and what we actually notice underscores the critical role of *attention*, arguably the most important organizational feature of our conscious life.[5] After all, there is no conscious perception without attention, which is indeed why we often fail to notice what is right in front of us, for example, unless our attention is in fact directed to it.[6] Visibility, in other words, is thus a function of not just strictly physical factors such as the object's size or distance from the viewer, but also the extent to which he or she actually attends to it.

As William James aptly summed it up, "[m]illions of items," indeed, "are present to my senses which never properly enter into my experience.... *My experience is what I ... attend to*."[7] To fully understand how we experience the world, we therefore clearly need to pay greater attention to the way we pay attention.

A most critical element in the process of attending is the mental act of *focusing*. "In attention," wrote John Dewey, "we focus the mind.... So the mind, instead of diffusing consciousness over all the elements presented to it, brings it all to bear upon some one selected point, which stands out with unusual brilliancy and distinctness."[8] As a "cognitive attitude," focusing therefore implies "an underlying preference for experiencing the world in a narrowed ... way."[9] In other words, it involves "disengag[ing] from a broader field of attention ... for the sake of ... focusing on a reduced number of stimuli."[10] As such, it inevitably entails a narrowing of our conscious awareness[11] by what Aldous Huxley so vividly described as the workings of some cerebral "reducing valve."[12] This is most evident, for example, in concentrative meditation, where we try to attain a single-focused mental state aptly known as "one-pointedness" by effectively restricting our awareness to a single source of stimulation (a particular "attention anchor"[13] such as a candle, stone, or mandala on which we fix our gaze while trying to suppress any stray thoughts; a special mantra we recite; or our breathing) and withdrawing it from anything that might distract us from focusing on that "anchor."[14]

Focusing, in short, implies *selective attention*, a particular mode of awareness encompassing only part of what actually impinges on our senses. The study of attention, to quote Paul Wachtel, is therefore "essentially the study of selectivity in perception and cognition."[15]

The fact that we actually notice only a few out of many potential perceptual stimuli[16] underscores attention's inherently exclusionary nature. As a process of selection, in other words, it

NOTICING AND IGNORING | 3

also implies exclusion.[17] When we attend to something, explained Dewey, we basically "fixate [its] mental content in the centre of the mind's activity, and allow all else to become dim and indistinct,"[18] effectively blocking much of what we can access through our sense organs out of our awareness.[19] As Wilhelm Dilthey described this,

> If I am looking out the window . . . the light of consciousness may well distribute itself evenly over the entire landscape. But as soon as I try to apprehend a single tree . . . in greater detail, the consciousness which I direct toward the rest of the landscape diminishes.[20]

A rather common metaphor for the process of attending, in fact, is that of a spotlight,[21] which, as so vividly exemplified in Figure 1.1, in selectively illuminating only a portion of the visual world separates that to which we attend from what we exclude from our awareness and effectively ignore:

> Attention may be compared to a beam of light in which the central brilliant part represents the focus surrounded by a less

FIGURE 1.1 The "spotlight" model of attention. Ted Leung Photography

intense fringe. Only the items located in the focus are distinctly perceived whereas we are less aware of the objects located in the fringe of attention.[22]

Attention, in other words, functions like a spotlight. Whatever lies within its focus is well noticed, whereas what remains outside it is effectively ignored.[23]

Indeed, by "concentrating the mind upon a small number of objects," observed Emile Durkheim, attention actually "blinds it to a greater number of others."[24] Or as Alfred Binet so aptly put it, focusing on a single point "increas[es] the intensity of that point so as to surround it with a zone of anaesthesia. Attention only increases the force of certain sensations in proportion as it attenuates others."[25]

In short, "attention is essentially a zero-sum game: If we pay more attention to one place, object, or event, we necessarily pay less attention to others."[26] Thus, as I focus my attention on the newspaper I am reading while eating my breakfast, for example, I inevitably pay less attention to the taste of what I am actually eating.[27]

Some fifteen years ago Daniel Simons and Christopher Chabris conducted a famous psychological experiment in which participants were shown a short video featuring two groups of people passing basketballs and were asked to count the number of passes made by one of those groups. Halfway through the video, a woman wearing a gorilla suit walked into the scene, faced the camera, thumped her chest, and walked off. Focusing their attention on the ball passes, however, about half of the participants did not even notice the "gorilla."[28]

As this now-classic experiment so spectacularly demonstrates, such *inattention*,[29] also referred to in such cases of visual imperception as "inattentional blindness,"[30] is actually the result of being attentive, yet to something else.[31] Indeed, ironically, it is usually when we attend to something in a highly focused manner that we also fail to notice other things around us. "[H]ow readily," observed Joseph Jastrow, "what, at one moment, is carefully fixed upon the charge of the attention, becomes lost in the background when other urgent claimants displace it."[32] Thus, while monitoring swimmers in a pool, lifeguards, for example,

> often fail to notice otherwise obvious events because they happen outside the[ir] immediate focus of attention . . . [E]ven though the[y] were "looking" right at the missed events, their attention was focused [elsewhere] . . . [E]ven the most vigilant observers suffer from [such] blindness. In fact . . . the more focused guards become . . . the more susceptible they are to failing to notice events that are outside of this intense focus of attention.[33]

Inattentional blindness thus occurs, in other words, not despite, but actually as a result of, being "focused."[34]

Inattention is manifested not only visually, however, as exemplified by auditory inattention, such as failing to hear certain sounds when we focus our attention on others,[35] as well as tactile inattention, which, as evidenced by various relaxation techniques,

> can easily be demonstrated by switching one's focus of attention to different locations on the body. For example, if you switch your focus of attention to your foot, you immediately become conscious of sensations arising from receptors in your foot that were non-existent a moment earlier.[36]

Indeed, as we shall see, the selective nature of our attention is evident not only in the organization of our sensory experience but also in the remarkably similar organization of the way we think about as well as remember things. It is likewise evident in the cognitive organization of our moral concerns, as any given set of moral considerations effectively "goes out of focus" whenever a competing one "comes into focus."[37] Given the striking similarity between the ways in which we focus our attention perceptually and conceptually,[38] we thus often fail to notice things that are "right in front of us" not only literally but also figuratively.

Furthermore, as evidenced by the choice of highly evocative titles like *Wide-Angle Vision* and *Peripheral Vision* for books about overcoming mental "blind spots" in management and business, or *Never Saw It Coming* for a study of the cognitive challenge of "envisioning" the worst,[39] we also invoke vision metaphorically in reference to non-sensory cognitive activity. Throughout this book I thus use visibility and invisibility as metaphors for *relevance* and *irrelevance*. And as exemplified by the case of camouflage, an in-depth study of invisibility thus constitutes, as we shall see, a perfect case study of the way we generally notice and ignore things.

As implied in the evidently symbiotic relation between attention and inattention, noticing also presupposes the act of *ignoring*. Given that it effectively involves not only taking in but also screening out information,[40] attention is thus indeed often associated with the metaphor of a filter,[41] which vividly captures the dialectical relations between perceptual passage and blockage. After all, notes Asia Friedman, filters "allo[w] selected elements to pass . . . while blocking others . . . [A]ll filters perform this function of 'straining' or 'sifting.' "[42] "Thinking in terms of filters," she adds, thus helps force us to consider "which features or details pass through and are attended and, perhaps more importantly, which are blocked by the filter and thus remain unnoticed."[43]

Studying how we notice things thus also presupposes studying how we effectively ignore others. As such, it calls for a fuller understanding of "the epistemology of ignorance," to paraphrase Charles W. Mills.[44]

The most ambitious effort to develop such an understanding has thus far revolved around the phenomenon of denial, famously identified by Anna Freud as a defense mechanism to which we often resort "in order not to become aware of some painful impression from without."[45] Such *unawareness* (either in its extreme form, as in the case of "psychic numbing" among the severely traumatized, or just in the form of "a continuous vigilance to *not* notice something,"[46] also known as "selective inattention")[47] is "a powerful survival technique when information is too dangerous to know."[48] In order to psychologically survive living in an abusive environment, for example, victims of intimate partner abuse thus often restrict their awareness of the abuse.[49] A somewhat similar tactic, ironically, likewise allows perpetrators only "a diminished awareness" of their victims' feelings.[50]

Inattention, however, need not necessarily manifest itself only in the form of actual avoidance. In fact, it quite often involves only an implicit absence of awareness. As such, it actually constitutes but a mere non-event, thereby implicitly posing a serious methodological problem for anyone who might venture to study it. The problem, of course, as Simons and Chabris indeed point out,

> is that we lack positive evidence for our lack of attention. . . . We are aware only of the . . . objects we do notice, not the ones we have missed. . . . It takes an experience like missing the chest-thumping gorilla . . . to show us how much of the world around us we must be missing.[51]

Being cognizant of this problem, I thus draw on the classic *figure-and-ground* model of perception, which most effectively captures the essence of the relations between what is perceptually present and absent when we attend to something in a focused manner. According to the Gestalt theory of perception, whether we notice something is largely a function of the way it is perceptually situated as a figure against its background-like surroundings. The relations between "figure" and "background," in other words, basically represent the relations between the attended and unattended parts of our phenomenal world.

As I argue in this book, although originally theorized within the specific context of visual perception, the figure-and-ground model is also applicable to non-visual (auditory, olfactory, gustatory, tactile) forms of perception as well as to altogether non-sensory modes of cognition. Indeed, throughout the book, I thus use "figure" and "background" to metaphorically represent the things we not only perceptually but also conceptually attend and inattend. By effectively examining the relations between the relevant and the irrelevant, I thus demonstrate that the fundamental principles underlying the act of sensory focusing also capture the essence of the more general act of *mental focusing*.⁵²

Unlike the Gestalt theorists, however, I am particularly careful not to essentialize the figure-like or background-like nature of what we only conventionally, after all, come to regard as figures or as backgrounds. Whereas the classic figure-and-ground model assumes that our attention is naturally directed to unmistakably figure-like entities that are inherently more perceptually prominent than the seemingly empty, unquestionably background-like spaces surrounding them, those entities are by no means inherently figure-like and those spaces by no means inherently background-like, and we only view them as such because we are habitually accustomed to do so. In fact, drawing on various artistic, intellectual, as well as explicitly political attempts to effectively reverse conventional attentional patterns by subversively directing our attention to what is conventionally unattended, thereby foregrounding what we habitually regard as background, I thus demonstrate that nothing is inherently figure-like or background-like.

Indeed, we also need to be careful not to essentialize the very distinction between figures and backgrounds, a distinction based on our ultimately conventional vision of seemingly discrete, freestanding entities that are somehow separable from their surroundings.⁵³ What we notice, after all, is not really detached from what we habitually ignore, and the contours we envision delineating the former and thereby separating it from the latter are in fact mere figments of our minds.

We likewise need to rethink our notions of relevance and irrelevance. What we consider "relevant" or "irrelevant," after all, is to a large extent merely a function of the way our attention is *socially*, and thus ultimately conventionally, delineated. Despite the fact that we often regard what we essentially ignore when we focus our attention on something as irrelevant or "extraneous," nothing is inherently irrelevant or extraneous.

Viewing something as inherently irrelevant, therefore, effectively essentializes our ultimately conventional notion of relevance by portraying it as if it were a strictly logical matter. Yet it is clearly not simply logic that makes us consider cockroaches morally irrelevant and compels jurors to disregard inadmissible evidence presented to them in court. Indeed, it is unmistakably social *norms* of attending that often determine what we consider relevant, and thereby notice, and what we complementarily consider irrelevant, and thereby ignore.

Yet, as is implicit in my discussion of the figure-and-ground model, this anti-essentialist critique of the limits of Logic is just as applicable to Nature. Admittedly, given the role of our brain as well as sense organs in the process of attending, whether or not we notice something is partly a function of our physiology as human beings, and most of the work on human attention thus far has indeed been done by biologically oriented psychologists and neuroscientists. Yet, while it has clearly advanced our understanding of the bio-cognitive hardware we use in order to notice things, such work has largely ignored the socio-cognitive software underlying the process of noticing and ignoring.

Whereas the similar manner in which brown pelicans focus their attention as they dive for fish is ultimately a function of their physiology, the strikingly similar manner in which structural engineers focus theirs when they inspect a building is not. By the same token, unlike the difference between eagles' and mosquitoes' ranges of vision, the difference between dentists' and astrophysicists' professional concerns (and therefore also sense of relevance) has little to do with any biological constraints on their ability to access the world through their sense organs.

What we notice and ignore, in short, is also a function of whether we are detectives or philosophers, Koreans or Americans, omnivores or vegans. In other words, *we notice and ignore things not only as human beings but also as social beings.* In order to study human attention, psychology and neuroscience are therefore clearly not enough. We also need a *sociology of attention*.[54]

A sociology of attention also highlights our often-shared and therefore ultimately collective sense of relevance and concern, thereby reminding us that *we actually notice and ignore things not only as individuals but also jointly, as parts of collectives.* As exemplified by the way various problems are collectively ignored, it thus also helps reveal our collective blind spots.

The present book is an attempt to uncover and explore the social underpinnings of human attention. To some extent, I contend, noticing and ignoring are *sociomental*[55] acts ultimately performed by members of particular communities with particular styles of attending, who focus their attention on particular slices of physical or mental reality while inattending others. As members of such communities, I show, we are *socialized* into culturally, subculturally (ideologically, professionally), and historically specific norms, conventions, and *traditions* of attending that actually determine what we come to regard as attention-worthy and what we effectively ignore.

Inevitably, the book reveals the remarkable extent to which we in fact reduce so much of what we can potentially experience both perceptually and conceptually to a mere "background" that we so casually ignore. Yet by at the same time uncovering the merely conventional nature of this process, it also helps remind us that such "background" need not actually always remain hidden in plain sight.

2

FIGURE AND BACKGROUND

> No field of view . . . is perceived all at one dead level. Some part of it will always tend to become "figural," and to be differentiated from the rest of the field which forms the "ground" to this "figure."
>
> MAGDALEN D. VERNON[1]

Whether we actually notice something, as I pointed out earlier, is to a large extent a function of the way it is perceptually situated as a *figure*, to which we focally attend, against some *background* (or simply "ground"), which we effectively ignore.[2] The significance of such "background" was already implicit in the works of James (in 1890), Jastrow (in 1906), as well as Edward Titchener (in 1910),[3] yet it was Edgar Rubin who placed the concepts "figure" and "background" in the very foreground of a theory of perception, explicitly titling the opening chapter of his groundbreaking 1915 doctoral dissertation "Figure and Ground."[4]

The distinction between "figure" and "background" represents the pronouncedly asymmetrical phenomenological distinction between the attended and unattended parts of our phenomenal world. In any given perceptual field, observed Rubin, the figure is always more attentionally prominent than the background.[5]

As etymologically implied in the fact that we effectively perceive it as being under[6] ("ground") or behind[7] ("background") the figure (as so vividly exemplified by the ancient Assyrian relief in Figure 2.1), the background is of only secondary importance relative to the figure it perceptually supports.[8] This is symbolically manifested in the fact that in portrait painting, for example, "background" elements such as the drapery were once painted

12 | HIDDEN IN PLAIN SIGHT

FIGURE 2.1 In marked contrast to the richly articulated figure, the background has no eye-catching features. The Trustees of the British Museum

only by the portraitist's assistant.[9] Indeed, as evident from young children's drawings, our earliest artistic representations actually consist of essentially backgroundless figures[10]—a rather telling pattern, further reinforced by commercial coloring books.[11]

In fact, we usually perceive the background as but an interspace,[12] often also characterized as a "negative" space.[13] Whereas figures are viewed as "things," backgrounds are perceived only residually, as the intermediate, hole-like spaces between them,[14] as exemplified by the way we conventionally envision urban space in terms of figure-like buildings separated from one another by stretches of open, "empty" space.[15]

Figures' thing-like quality is visually underscored by the perceived non-thing-like quality of their backgrounds.[16] Not only are the sculpted figures in the ancient Greek relief in Figure 2.2, for example, raised (practically as well as symbolically) above their background-like surroundings, they are also

FIGURE 2.2 The sculpted figures of the Three Graces are raised physically as well as symbolically above their backgrounded surroundings. RMN-Grand Palais/Art Resource, NY. Photo: Hervè Lewandowski

articulated much more vividly.[17] Indeed, one of the foremost characteristics of the background is the fact that it is not as richly (nor, for that matter, as clearly)[18] articulated as the figure it perceptually supports.[19] As exemplified by Figure 2.1, whereas figures are usually articulated in great detail, backgrounds merely provide the effectively empty[20] context within which they are perceptually situated and have practically no "eye-catching" features.[21]

That explains why we often fail to notice the details of, as well as changes in, the background-like parts of our perceptual field,[22] whose texture usually seems rather indistinct.[23] It also explains the traditional distinction between the different Oscars respectively awarded by the Academy of Motion Picture Arts and Sciences every year to the performances by the "Best Actor in a Leading Role" and the "Best Actor in a Supporting Role," one of which is clearly expected to attract the audience's attention much more prominently than the other.

As one might expect, differences in attentional prominence entail differences in both perceptibility and recognizability. In marked contrast to the figure, the background is viewed as effectively shapeless[24] as a result of being perceived as lacking a well-delineated contour that would provide it with a thing-like character.[25] Shape is a product of delineation,[26] yet as Rubin so astutely observed, we tend to perceive the contour actually adjoining the figure and the background as bounding the former but not the latter.[27] The pronouncedly asymmetrical relation between the figure-like and background-like parts of a given perceptual field, in other words, essentially boils down to "the different roles the contour plays for them. Properly speaking, the contour belongs entirely to the figure and has no significance for the ground."[28]

We thus tend to perceive coastlines, for instance, as delineating the land but not the sea around it,[29] which is indeed why we seem to perceive "the famous boot of Italy and the outstretched palm of Greece," for example, yet fail to see "the form of the Adriatic Sea between them. If asked could we say off-hand what object it resembles? Surely not; in fact we are totally unfamiliar with it."[30] That, of course, applies just as well to the unoccupied space in our living rooms[31] and to the patches of sky we see between buildings or trees. As we can see in Figure 2.3, for example, whereas the buildings themselves seem to have shapes, we tend to perceive the sky between them as effectively shapeless,[32] which is indeed just as true for almost all shapes that . . . give the impression of [a] background. Of all shapes projected into the eye, we can usually only really *see* those that give the impression of figures [yet when] an object creates the impression of an intermediate space, it is as though it has been made to disappear by magic even though it is lying in plain view before our eyes.[33]

In short, whereas figures are perceived as having well-delineated boundaries, backgrounds are viewed as essentially unbounded.[34] Thus, whereas human settlements, for example, are often envisioned as being bounded, the wilderness surrounding them is not.[35] And since it is their boundaries that actually provide figures with their perceived shape, backgrounds' perceived boundlessness indeed makes them seem shapeless. Thus, whereas

FIGURE 2.3 In marked contrast to the figure-like buildings, the background-like patches of sky between them are perceived as shapeless. Photo taken by the author

16 | HIDDEN IN PLAIN SIGHT

FIGURE 2.4 We can alternately focus on either the faces or the vase but not on both of them at once. Image courtesy of Ilanit Palmon

"[o]wing to the shape the figure derives from the contour, it appears as a self-contained unit, detached from the ground,"[36] and therefore seems recognizable, the background's perceived boundlessness makes it seem invisible.

This, indeed, is most compellingly exemplified by the so-called "Rubin vase," the famous "vase-faces" image that, since having been featured by Rubin in his doctoral dissertation,[37] has certainly come to emblematize the figure-and-ground model of perception.[38] As we can see in Figure 2.4, the shapes of the vase and the faces "cannot be simultaneously observed: only one shape or the other will be seen at any one moment in time."[39] We can thus perceive either the vase or the faces, but not both of them at once, together. As we focus our attention on either of them, the other thus seems effectively shapeless and, as such, remains practically invisible.

Originally conceptualized specifically within the context of visual perception, the figure-and-ground model is nevertheless applicable to non-visual modes of perceiving as well[40] (and what is true of visibility and invisibility, as we shall see, is also true of perceptibility and imperceptibility in general). When smelling or tasting a curry, for instance, I can thus pick out the distinct figure-like odor[41] or flavor of the cumin or garlic in it from its background-like olfactory or gustatory surroundings. Moving one's hand over a stiff brush, one may likewise experience its bristle points as "tactual figures" and the spaces between them as a "tactual ground."[42]

Yet the applicability of the figure-and-ground model beyond the visual domain is most strikingly evident in our ability to differentiate auditory figures from the acoustic background in which they are perceptually embedded,[43] as exemplified by the way we habitually parse continuous spoken input into distinct words. Although the background-like blank spaces we conventionally insert between figure-like written words to separate them from one another are not always present in actual speech,[44] we nevertheless need to mentally "hear" them in order to be able to distinguish "nitrate" from "night rate" and "no notion" from "known ocean"[45] or make sense of the effectively continuous auditory stream "perhapstheyshouldhavetrieditearlier,"[46] and the figure-and-ground model certainly provides the most compelling account of our ability to do so.

Even more spectacular, however, is the way we manage to habitually assign the plethora of concurrent sounds entering our head to their essentially distinct, separate sources:

> Suppose you are in an environment where there is sound coming from a television set, sound coming from a child playing, and sound coming from a person talking. You are able to separate these sounds into recognizable streams; they do not blend together as one sound conglomerate.[47]

Our ability to do so, however, presupposes a most intriguing mental process identified in 1971 by Albert Bregman as "auditory stream segregation," or simply "streaming,"

> in which a single ... sequence of tones seems to "break up" perceptually into two or more parallel sequences. ... We call this phenomenon primary auditory stream formation. A stream may be defined as a sequence of auditory events [that are] being segregated perceptually from other co-occurring auditory events.[48]

The process of parsing an acoustically complex situation into several distinct (and thus separate) auditory "streams" is remarkably analogous, of course, to the mental process of carving up visually complex perceptual fields into separate figure-like and background-like sections. And just as in the case of the faces and the vase, we can actually focus our attention on only one stream at a time. As we focally attend to any given stream, all the other acoustic information is thus effectively relegated to the background.[49]

As one might expect, when a given sequence of sounds is bound into a single identifiable stream, it is easier to attend to it and thus to mentally "follow" it over time.[50] It is "streaming," therefore, that allows us to follow what somebody is saying to us despite the constant sound of traffic in the street, listen to the radio while children are noisily playing around us, and pick out any particular one of the many conversations simultaneously going on around us at a crowded party or restaurant.[51] And it basically involves focusing our attention on only one of the many potential auditory figures available while "tuning out" all the rest as mere background.

Consider also in this regard the conventional distinction between "melody" and mere "accompaniment," which we come to perceive as but "the acoustic background ... against which melody stands out in relief as an individual foreground figure."[52] As we listen to music, we thus usually focus our attention on (and thereby mentally "follow") one supposedly "melodic" line and essentially regard all other simultaneously produced sounds as mere "background" embellishment.[53] As they transition from "merely" accompanying to actually playing a featured solo, big-band jazz musicians indeed often stand up to visually mark the dramatic shift between their perceived roles as background-like and figure-like acoustic objects.[54]

As the acoustic equivalent of the distinction between lead and supporting acting, the distinction between melody and accompaniment is pronouncedly hierarchical, with the former considered attentionally dominant while the latter is perceived as playing a merely secondary role. Accompaniment thus constitutes "that part of a musical continuum generally regarded as providing support for, or the background to, a more prominent strand in the same music."[55] Consider, for example, in this regard the traditional "background" role of the double bass and the bass guitar in classical, jazz, and rock music.[56] Though absolutely captivating, Bob Cranshaw's bass line on the Horace Silver Quintet's *Bonita* and Paul McCartney's on the Beatles' *With A Little Help from My Friends*, for instance, were nevertheless designed to be perceived as merely supporting those songs' "main" melody lines.

The accompaniment's essentially supportive role is also manifested rhetorically in the way we use the word "backing" (as in "backing vocals" and "backing track") to characterize the sounds we conventionally regard as but the acoustic background to those produced by the "lead" vocals or instruments.[57] In fact, in live jazz and rock performances, effectively participating in forming a spatial structure that visually embodies the hierarchical relations between figure-like and background-like acoustic objects, drummers, for instance, are often placed literally behind the featured soloist or lead singer (who are indeed also known as the "frontline" musicians), thereby visually reinforcing their perceived role as merely "backing" them up.

In fact, there is even an entire musical genre effectively conceived as but mere background. Often played in supermarkets, elevators, and public lavatories, "background music" is clearly not meant to actually attract anyone's attention. As a sonic form of wallpaper,[58] it represents our auditory equivalent of peripheral vision ("peripheral hearing").[59] Indeed, it is designed to be unobtrusive, "loud enough to give you an idea of what's playing but low enough to keep it on the brain's back burner," something we should actually "hear without listening."[60]

A perfect example of such "background" music is the cinematic soundtrack, specifically designed to be "subtle enough not to overscore the screen action," that is, "catchy yet innocuous enough not to detract from the story."[61] As sometime film composer Aaron Copland glumly observed, it is "the most ungrateful kind of music for a composer to write. Since it's music behind, or underneath, the word, the audience . . . possibly won't even be aware of its existence."[62]

RELEVANCE AND IRRELEVANCE

Although originally conceptualized specifically within the context of sensory perception, the figure-and-ground model is nevertheless applicable to non-sensory modes of cognition as well. Given the considerable affinity between the ways in which we focus our perceptual and conceptual attention,[63] it is hardly surprising, therefore, that the basic principles underlying the processes of visual and auditory focusing also capture the essence of the process of *mental focusing*.[64] The way we distinguish "lead" actors from merely "supporting" ones when watching a film is thus remarkably similar to the way we distinguish "major" from "minor" characters when reading a novel.

In fact, the concepts "figure" and "background" help us to phenomenologically describe the things we not only physically but also mentally notice and ignore,[65] as exemplified by examining the way "justice" and "care" function as alternative perspectives guiding our "moral attention."[66] Whereas from a justice perspective "the self as moral agent stands as the figure against a ground of social relationships," from a care perspective it is those relationships that in fact become the figure.[67] By the same token, one can also describe the process of remembering and forgetting in terms of the way we differentiate figure-like "eventful" (and therefore memorable) parts of the past from essentially background-like "uneventful" ones which are effectively relegated to oblivion.[68]

Indeed, figure (and, for that matter, foreground) and background actually constitute metaphors for relevance and irrelevance, as manifested in statements such as "In conversations with worried donors and aides, [Romney] repeatedly returns to the simple math of the nomination process: he needs 1,144 delegates. The rest, he says, is background noise."[69] Thus, for example, "in any speaking situation," note linguists,

> some parts of what is said are more relevant than others. That part of a discourse which does not . . . crucially contribute to the speaker's goal, but which merely assists . . . is referred to as BACKGROUND. By contrast, the material which supplies the main points of the discourse is known as FOREGROUND.[70]

By the same token, in narrative analysis, "foreground" represents "the main story-line" whereas "background" represents the merely "supportive" material.[71]

Consider also what we conventionally regard as "out-of-frame,"[72] which, by contrast to that which is "framed"[73] and to which we are therefore expected to attend, we are supposed to relegate to the "side tracks"[74] of our awareness and effectively ignore. As Gregory Bateson, who pioneered the study of mental framing, described the frame both literally and symbolically separating the figure-like, artistically relevant space of a painting from the effectively background-like, artistically irrelevant space of the wall surrounding it,

> The frame around a picture . . . as a message intended to . . . organize the perception of the viewer, says, "Attend to what is within and do not attend to what is outside." . . . Perception of the ground must be positively inhibited and perception of the figure (in this case the picture) must be positively enhanced.[75]

Such attentional split is likewise manifested in the way concertgoers may sometimes not notice musicians' "background" activities (turning the pages of one's sheet music, wiping the spittle from one's horn) that are not considered part of the artistic performance in which they are nevertheless visually embedded.[76]

By the same token, contestants' highly visible features such as their height or the color of their hair are nevertheless considered irrelevant, and therefore effectively ignored, in Scrabble tournaments and spelling bees. And when we play checkers and a particular piece is missing, we may quite casually replace it with a coin, essentially disregarding the latter's monetary value as utterly irrelevant to the game.

THE MARKED AND THE UNMARKED

Furthermore, figure and background in fact constitute metaphors not only for the relevant and the irrelevant but also for the "marked" and the "unmarked" parts of our phenomenal world.[77] The distinction between them therefore represents the fundamental distinction between the remarkable (and thus *noteworthy*) and the unremarkable (and thus *unnoteworthy*), as exemplified by the way we conventionally distinguish holidays from "ordinary" days,[78] having a Southern accent from having "no accent," and atypical from "normal"[79] behavior.

As so vividly exemplified in the case of visual figures and backgrounds, the relation between the marked and the unmarked is fundamentally asymmetrical. In sharp contrast to marked mental constructs ("spicy," "contagious," "fiction"), unmarked ones are residually articulated as that which is *not* marked ("non-spicy," "non-contagious," "non-fiction").[80] Indeed, as implied in the fact that only one of the days of the Jewish week has a distinctive name (the Sabbath) while the other six are simply designated by ordinal numbers indicating their distance from it,[81] whereas marked mental constructs are usually named, unmarkedness often manifests itself in namelessness.

In other words, whereas the marked parts of our phenomenal world are explicitly articulated (thereby becoming special objects of attention), the unmarked ones are only implicit. Thus, for example, in America, whereas blackness, which is socially marked, implies the presence of African ancestors, whiteness, which is socially unmarked, implies their absence rather than the presence of European ones. Whereas having even a single African ancestor therefore often suffices to make Americans "black," having several European ancestors does not always suffice to make them "white."[82]

Such fundamental asymmetry is largely a function of relative size. As so vividly evident in the case of visual figures and backgrounds,[83] the marked is usually more narrowly specified than the unmarked.[84] As exemplified by the ratio of gendered parts of our body to generic ones, for instance,[85] *what we explicitly notice is proportionally smaller than what we implicitly ignore.*

Essentially considered unremarkable and therefore unnoteworthy, the unmarked parts of our phenomenal world are usually relegated to "the background." As we shall now see, that certainly affects what we come to notice as well as what we fail to notice.

3

SEARCHING AND HIDING

> A black cat in a coal bin is difficult to see. A black cat on a snow bank is hard to miss.
> —CARL PURCELL[1]

SPOTTING

To further appreciate the figure-and-background structure of our phenomenal world, let us also examine the cognitive dynamics of *searching*, a mental process aimed at *spotting* figure-like "targets" by essentially differentiating them from the background-like surroundings in which they are perceptually embedded.[2] In theorizing the phenomenology of this process, it may be useful, for example, to invoke the notion of a figure-like "signal" perceptually embedded in background "noise."[3] The higher the ratio of the intensity of this signal to the level of its surrounding noise, of course, the easier it is to spot it. It may also be useful to invoke the metaphor of a searchlight.[4] Effectively designed to help us *scan* large bodies of information in search of specific targets, both the back-of-the-book index and the search engine, for instance, ultimately presuppose the kind of mental "illumination"[5] it figuratively implies.

As we scan our perceptual field when searching for something, we basically disregard anything we consider irrelevant. As Ulric Neisser described the mental process of scanning a phone book when looking up the number of some Mr. Smith who lives on Fifteenth Street, for example, there are usually, as one would expect,

> a lot of Smiths, but it does not take long to skim down the column to the correct address; the irrelevant addresses are passed over so

quickly that they seem blurred. In a sense they are not seen at all. In much the same way one sees, but does not see, dozens of hurrying figures when trying to locate a friend in a busy air terminal.[6]

By the same token, if you were looking for a particular brand of shampoo in the supermarket,

> you would actually "set" your attention . . . to screen out bottles that look different from [that] brand. . . . You wouldn't have to carefully scrutinize every bottle on the shelf. . . . You'd let your eyes scan the shelf, paying little attention to most of the shampoo, zeroing in quickly on the one you wanted.[7]

That implies, of course, that as we begin a search we already have a certain mental "search image" guiding it. Thus, when looking for particular mushrooms, for example, there is "a little template that you carry in your mind." In fact, if "you just walk around with your head down . . . you won't see anything."[8]

Effectively mirroring the process of searching for something is the process of trying to "capture" someone's attention.[9] Complementing shoppers' efforts to find that particular brand of shampoo in the supermarket, for example, are the store manager's efforts to display it in such a way that they would in fact be able to spot it, not to mention advertisers' efforts to get them to look for it in the first place.

In order to attract attention (thereby figuratively "catching" someone's eye), we sometimes dress colorfully so as to visually "stand out from the crowd." And in an effort to make the "structure of relevance"[10] we wish to impose on texts we read visually salient so as to capture our own or someone else's future attention, we likewise mark certain passages with a bright highlighter pen (just as we underline or italicize certain words when writing or typing):

> A quite general class of cognitive practices consists of methods for highlighting [a] perceptual field so that *relevant phenomena are made salient*. . . . Faced with . . . a dense perceptual field,

> workers [often] highlight their documents with colored markers
> ... [They thus] tailor the document so that [certain] parts of it
> ... are made salient.[11]

By the same token, our ability to recognize a familiar tune is partly a function of how effectively we can differentiate it from the other sounds in which it is acoustically embedded.[12] And when listening to polyphonic music, we can much more easily follow two concurrent melodies when they are not in the same pitch range, when they have noticeably different rhythms, or when they are played by different instruments, and are therefore more easily differentiated from each other.[13]

Differentiation, in short, is a precondition for awareness.[14] "[Our] perceptual mechanism has evolved to cope with a differentiated field, and, in the absence of differentiation, there is a temporary breakdown of the mechanism,"[15] as so dramatically exemplified by the "ganzfeld" effect generated by perceptually homogeneous surrounding environments (a dense fog cloud, an MRI tunnel, Arctic whiteouts), when it is practically impossible to perceive any "thing" at all or to even tell whether one's eyes are actually open or closed.[16]

Yet it is not simply differentiation, but actual *contrast*, that makes things more discriminable and therefore more likely to be spotted. A sharp contrast between a figure and its surrounding background makes it easier to notice its contours and thereby perceive its shape,[17] thus helping us to actually notice it. In fact, when a figure is perceptually embedded in a contrasting background, it seems to almost literally "pop out" of it.[18] "The degree of . . . contrast between an object and its background is what makes it visible. . . . A black cat on a snow bank is hard to miss."[19]

That explains the importance of high contrast in biomedical imaging, as exemplified by the way magnetic resonance imaging (MRI), for instance, helps detect problems that might not even be noticeable in low-contrast images, where objects' contours are not as sharply delineated. By the same token, while scanning video footage in an effort to identify the 2013 Boston Marathon bombers, it was a striking behavioral contrast that actually helped the authorities spot them: "According to officials, when the blasts went off, most people fled in panic, but these two did not and instead walked away slowly, almost casually."[20]

IN THE BACKGROUND

As one might expect, things tend to become less *noticeable* when the general conditions that facilitate spotting them are not present. The very aspects of the relation between a figure and its surrounding background that make it easier to spot it (such as the existence of a sharp contrast between them), in other words, thereby make it harder to spot when they are absent or reversed.

As so vividly exemplified by Edgar Allan Poe's famous "purloined letter"[21] (or by "Verbal" Kint, Kevin Spacey's character in the film *The Usual Suspects*, for that matter), things are least "spottable" (and thus most *unnoticeable*) when they are either literally or figuratively "in the background." As such, they are usually *taken for granted* and thereby tacitly ignored.[22]

Thus, as a result of cognitive habituation, for example, we rarely notice that which is constantly around us and therefore lacks the contrastive quality that would have helped us spot it. As Ludwig Wittgenstein so wisely observed, it is in fact their very familiarity that often makes things unnoticeable. "One is unable to notice something," in other words, precisely "because it is always before one's eyes."[23]

When getting dressed, for example, we can thus

> feel our clothes on our skin, but then the feeling fades.... If the senses are exposed to a continuing stimulus, habituation soon occurs.... This is why we do not feel the clothes we are wearing and become aware of them only if we change or adjust them.[24]

That is also true of the constant sound of our breathing, our own body odor, and the persistent pressure of the shoes we are wearing against our feet.[25] As ordinary, nearly ever-present and thus experientially unmarked components of our phenomenal world, they lack remarkability and therefore usually escape our attention.

Our cognitive habituation to what we consider unmarked is the main reason, for example, why it has taken sociologists longer to focus their scholarly attention on masculinity, adulthood, and well-being than on femininity, adolescence, and illness.[26] It also explains the fundamental difference between locals' and visitors' attentional patterns, as exemplified by the way many New Yorkers manage to effectively "tune out" the sound of the police-car sirens which for many tourists is such a distinctive feature of their city's soundscape. In fact, locals sometimes even fail to notice the very attractions that bring tourists to their town. "The tourist gaze," after all, is specifically "directed to features of [the] townscape which separate them off from everyday experience. Such aspects are viewed because they are taken to be in some sense out of the ordinary,"[27] which is precisely why locals, for whom they are part of their ordinary, ever-present and therefore experientially unmarked experience, may not even notice them.

Some objects, however, are perceptually relegated to "the background" simply because they are considered socially irrelevant, such as mere "bystanders," whom we usually notice only when they actually force themselves into our awareness, as when a total stranger butts into a conversation between a couple standing in line.[28] Like "extras" in films, such *background persons*[29] are often considered invisible.[30] Effectively regarded as "nonpersons,"[31] they may not "gain even the minimum attention required to feel that [their] presence [is] acknowledged."[32] According to Erving Goffman, who pioneered the study of *social invisibility*, they are treated as though they are not present, and are even expected to "act so as to maximally encourage the fiction that they aren't."[33] Such *attentional marginalization* is usually associated with being of low status,[34] as exemplified by the way we often ignore cleaning personnel, beggars, as well as children. Thus, in the film *Maid in Manhattan*, after soon-to-be senatorial candidate Ralph Fiennes falls in love with Jennifer Lopez, whom, unbeknownst to him, he actually first "met" when he was about to relieve himself in his luxurious hotel bathroom which she was in the midst of cleaning, she indignantly reminds him, "You think you would have taken a second look at me if you knew I was the maid? . . . I am just invisible. . . . The first time you saw me I was cleaning your bathroom floor, only you didn't see me."

For precisely the reason that it hinders our ability to spot things, however, their *inconspicuousness* also helps us when we try to hide them. Indeed, as a process ultimately antithetical to searching, *hiding* actually involves reversing the perceptual conditions necessary for a successful search. Given the fact that objects are more noticeable when they move against a fixed background, for example, one can easily understand why prey animals often "freeze" when trying to hide from a predator.

As exemplified by wanted criminals, members of terrorist "sleeper cells," ex-convicts, as well as incognito celebrities, inconspicuousness is basically achieved by "keeping a low profile" and effectively "hiding in the crowd." Thus, in the aftermath of the Boston Marathon bombings, for example, "[f]or more than three days—from the time of the explosions . . . until the F.B.I. released their photographs," it was indeed "the very ordinariness of their activities [that] let the [Tsarnaev] brothers hide in plain sight."[35] By the same token, in the Nazi concentration camps, inmates often tried not to become conspicuous, and to

> make [themselves] invisible. . . . At roll call, one had to disappear somewhere in the middle rows. . . . In the marching column, one needed to get oneself a position roughly in the middle. The maxim was: avoid the exposed outer edges. . . . If one crossed the field of vision of the supervisors, one had to move at a normal speed . . . not too fast, not too slow. . . . Unnoticed, one was in no danger of being selected as a victim of some punitive operation.[36]

It is the ability to "fit in" so as to *not stand out*, after all, that allows one to "pass." By downplaying one's markedness while playing up one's ordinariness,[37] one can thus stay "in the background" and remain effectively invisible.

As one might expect, their inconspicuousness makes background persons ideal candidates for spying. Robert Baden-Powell, who as a British intelligence officer in the late nineteenth century managed to produce detailed drawings of enemy military installations in Dalmatia while posing as a harmless butterfly collector,[38] and Elyesa Bazna ("Cicero"), who used his highly inconspicuous position as a valet to the British ambassador to Turkey to become one of Germany's most valuable spies in World War II, are perfect examples. So, indeed, are young children, who, often unnoticed by adults, can actually become quite useful in interfamilial espionage.[39]

BACKGROUND MATCHING

While a black cat on a snow bank may indeed be hard to miss, a black cat in a coal bin is actually difficult to see.[40] By the same token, in fact, it is probably easier to find a needle in a haystack than in a bin full of other needles.[41]

Indeed, the more something resembles its surroundings, the harder it is to spot.[42] "[B]ecause they have more features in common with the surround and therefore blend into it," embedded targets are generally "harder to find than nonembedded targets that stand out from [it]."[43] A figure, in short, tends to become conspicuous when it contrasts with the pattern of the background in which it is perceptually embedded, yet inconspicuous when it conforms to it.[44]

In the latter case, that is basically a result of the perceptual dissolution of its contours and therefore also its overall shape. The greater the resemblance between a figure and its surrounding background, the less noticeable its contours and the more difficult it is therefore to perceive it as a distinct entity.[45]

This is perhaps most vividly exemplified by the instrument we use to detect color-vision deficiencies. Originally designed in 1917 by Shinobu Ishihara, our standard screening test for "color blindness" consists of a set of plates featuring various assortments of dots of varying sizes, colors, and degrees of saturation and brightness specifically arranged so that similarly colored dots form a figure-like numeral surrounded by differently colored ones, with the colors of the figure and its immediate background being ones that people with anomalous color vision cannot actually differentiate from each other. Thus, whereas people with normal color vision can easily spot the numeral "74" in Plate 1, "color-blind" persons who cannot differentiate greenish from reddish hues, for example, cannot perceptually separate this greenish figure from the reddish background in which it is visually embedded and are therefore effectively unable to see it.[46]

In the absence of a sharp contrast that would help differentiate it from its immediate surroundings, in short, a might-be perceptual object is thus unlikely to become one. Failure to situate an object against a sharply contrasting background, in other words, makes it unlikely to be noticed.[47]

Indeed, as so dramatically exemplified in the way we manage to cosmetically conceal scars, burns, birthmarks, bruises, acne, and skin discoloration by making them appear visually continuous with their surroundings, by deliberately making a figure match (and thereby effectively blend into) its background we can actually decrease its perceptibility.[48] Noticeability, in short, can thus be reduced by simply reversing the conditions necessary for a successful search. After all, for the very same reason that the target must look different from its surrounding background in order to be more conspicuous, it must resemble it in order to remain inconspicuous.[49] And indeed, whereas spotting presupposes the ability to differentiate a figure from its surrounding background, *background-matching camouflage*[50] is essentially based on blurring the very distinction between them.

Camouflage, of course, "implies a seeing eye from which to hide."[51] And indeed, background matching manifests itself perhaps most vividly in coloration patterns that help make animals effectively invisible, as in the countless cases of prey coloration that so perfectly resembles the prey's actual habitat and is the reason why chameleons, for example, often change their color when moving between different habitats.

It was Erasmus Darwin, Charles's grandfather, who first observed that animals' colors "seem adapted to their purpos[e] of concealing themselves . . . to avoid danger," and that birds' eggs are likewise protectively designed "to resemble the colour of the adjacent objects"[52]—a most insightful observation later reiterated by Alfred Russel Wallace:

> The fact that first strikes us in our examination of the colours of animals . . . is the close relation that exists between these colours and the general environment. Thus, white prevails among arctic animals; yellow or brown in desert species; while green is only a common colour in tropical ever-green forests.[53]

The reason for this was explicitly articulated by Edward Poulton, who pointed out that an animal's colors, markings, and patterns help it "harmonise with the general artistic effect of its surroundings" specifically *"so that it does not attract attention."*[54]

Yet it was not a natural scientist like Darwin, Wallace, and Poulton but rather an artist, Abbott Thayer, who first developed a full-fledged attention-based theory of camouflage. As he and his son Gerald announced in their 1909 book *Concealing-Coloration in the Animal Kingdom*, "'Protective coloration' . . . has waited for an artist . . . to perceive that the many animals of supposed 'conspicuous' attire are almost all colored and marked in the way most potent to conceal them."[55] Animals' coloration patterns, in other words, are specifically designed to enhance their *in*conspicuousness:

> All the patterns and brilliant colors on the animal kingdom, instead of making their wearers conspicuous, are, on the contrary, pure concealing coloration, being the actual color notes of the scene in which the wearer lives, so that *he really is nature's utmost picture of his background.*[56]

In fact, argued Thayer, provocatively featuring a painting of a peacock, the very symbol of conspicuousness, blending in with its leafy surroundings, we usually see such seemingly conspicuous animals situated against the "wrong" background.[57] Indeed, as Frank Beddard had already pointed out, in their natural habitat their coloration actually matches their surroundings:

> Most persons who had only seen these animals in the Zoological Gardens would be inclined to look upon the giraffe, the zebra, and the jaguar as among the most conspicuously coloured of the Mammalia; and yet we are assured, by those who have seen them in their native countries, that they are most difficult to detect.[58]

"*Concealing-coloration*," concluded Thayer, thus means primarily "*coloration that matches the background.*"[59] Its underlying principle, which is basically that of "painting on the surface of the creature itself a picture representing as closely as may be [its] background,"[60] is the idea that a predator would "think that he sees background, when he is really seeing a part of the animal."[61] A bird, for example, is thus colored as "such part of the scene . . . as the eyes that he most needs to avoid would see him against."[62] Indeed, suggested Thayer, invoking the possible didactic use of cutouts,[63]

> [g]o to . . . where antelopes . . . come to drink, and with your eyes a foot from the ground (crouching lions' eyes height) study through an antelope-shaped hole in a card the [view that] would form these beasts' background to a crouching feline. What you discover is that . . . those color notes which are almost always present in the lion's view of the antelope, are fully repeated on the part of him this enemy is surest to see against such . . . background.[64]

Yet as Darwin, Poulton, Wallace, and Thayer all realized, camouflage not only helps prey animals hide from predators but also allows the latter to approach the former unseen.[65] In other words, it serves "to conceal the herbivorous species from their enemies" while at the same time also "enabling carnivorous animals to approach their prey unperceived."[66] The lynx's coloration, for example,

> just as much increases his dangerousness to the hare, as that of the hare adds to the lynx's difficulty in catching him . . . [The] *inconspicuousness* of the predator causes him to be less avoided by the animal he preys on, while for the prey it means a minimizing of the stimulus he gives to his enemy's rapacity . . . [B]oth sides profit by showing as indistinctly as possible, so that the rapacious animal is harder to dodge, and the prey a fainter target to strike at.[67]

The same dialectic, of course, is at work in human camouflage as well. While camouflage headdresses and ghillie suits, for example, help conceal snipers, similar forms of background matching are designed to protect potential targets. Such protection may not have seemed necessary when warriors fought hand-to-hand (and indeed sometimes even dressed somewhat flamboyantly to intimidate their adversaries), but it certainly became quite critical with the advent of long-range weapons such as the rifle, as in fact evidenced by the use of gray and green uniforms by some British and Austrian units in the eighteenth century and the adoption of khaki as the standard color of British military uniforms by 1902.[68] Gray, olive, and khaki uniforms were soon also adopted by the American, Japanese, Italian, Russian, German, and Belgian armies, and by 1915 even the French army finally gave up its traditional colorful uniforms, adopting instead drab ones to allow soldiers to better blend in with their surroundings.[69]

It was also during World War I, in the context of efforts to conceal military equipment, that the word *camouflage* was actually introduced. The idea of making fortifications "indistinguishable from the ground in which they stand" (instead of "batteries conspicuously frowning . . . with tiers of guns rivalling rows of targets in . . . clearness of definition") had already been raised in the 1880s,[70] but the need to conceal military equipment gained special urgency with the advent of aerial reconnaissance photography.[71]

It was in order to address that need that the French army invited the painter Lucien-Victor Guirand de Scévola to establish its *Section de Camouflage*, and by 1915 this special unit effectively made up of painters, sculptors, and set designers emblematically represented by the figure of a chameleon[72] began painting tanks, heavy artillery, and armored personnel carriers to match their surroundings (as well as covering them with nets splashed with earth colors), thereby blurring their contours in an effort to make them seem invisible.[73] Such optical tactics were soon also adopted by the British, Italian, Belgian, German, and American armies,[74] and by the end of World War I camouflage had already become an integral part of modern warfare.

Military camouflage has since then clearly also inspired a number of modern artists (Holger Trülzsch, Rachel Perry Welty, Désirée Palmen, Laurent La Gamba)[75] to experiment with invisibility by painting themselves as well as their models to actually match their surroundings. As perhaps most spectacularly exemplified by Liu Bolin's "camouflage art" (see, for example, Plate 2),[76] such works represent an epistemically subversive effort to optically eliminate a person's contours and thereby effectively obliterate his or her status as a distinct object of perception.[77]

Note, however, that efforts to blur the distinction between figure and background are by no means confined to visual perception. Like some insects,[78] hunters, for example, often try to elude the attention of their anticipated prey by effectively matching their olfactory surroundings (such as by storing their hunting gear along with leaves, twigs, and dirt from their hunting grounds), thereby downplaying their figural status and essentially becoming "smell invisible."[79]

There is a particular form of background-matching camouflage known as *steganography*, in which actual "signals" are purposely designed to be mistaken for mere "noise" and thereby effectively ignored. One can thus send a covert message in such a way that no one apart from oneself and one's intended audience (who are in fact usually alerted to expect it) even suspects its existence. Whereas in cryptography only the content of the hidden message is concealed, in steganography its very existence is concealed as well.

Steganography is, in short, the practice of hiding signals by embedding them within what seems like mere noise. Effectively presented as mere "background" and therefore as largely irrelevant, a covert signal can thus be overtly displayed without attracting any attention. As an ordinary-seeming object that looks like an integral part of one's backyard, a fake rock, for example, is for all practical purposes invisible and thereby constitutes a perfect place in which to hide a key.

That, indeed, is how graffiti artist and political prankster "Banksy" actually managed to surreptitiously smuggle a miniature model of a fighter jet with missiles slung under its wings into New York's American Museum of Natural History in the midst of America's Iraq War. Facetiously attaching it to a glass-encased beetle with the pseudo-scientific caption "Withus Oragainstus"[80] (thereby subversively referencing George W. Bush's famous slogan "You're either with us or against us"), he thus artfully managed to make it look like an integral part of the museum's Hall of Biodiversity.

Such was also the case with the songs (as well as quilt patterns)[81] often used in the nineteenth century by the Underground Railroad to provide American runaway slaves with actual escape plans:

> [S]ome slaves communicated their intentions of escape through songs whose words contained secret messages. In coded spirituals, slaves expressed . . . plans and directions for the journey North. . . . "Follow the Drinking Gourd" was a secret metaphor for the "Big Dipper," which served as a directional to the North [while] "Let Us Break Bread Together" . . . told slaves to meet before sunrise on the east side of the slave quarters to plan an escape.[82]

In a classic steganographic manner, slaves could thus *openly send secret messages* containing specific escape plans by essentially repeating them aloud while their captors were totally unaware of what was actually being secretly communicated through the very songs they were hearing.[83]

By the same token, while a prisoner of war in North Vietnam, U.S. Navy pilot Jeremiah Denton used the occasion of being forced to participate in a 1966 television interview as an opportunity to quite ingeniously "smuggle" out the information that he was being tortured by his captors. Covertly targeting American intelligence officers as his intended audience, he thus actually "spelled out" the Morse code equivalents of the letters of the word *torture* in quick (for dots) and slow (for dashes) eye blinks.[84] The very possibility of such "openly covert" communication was indeed the reason why in the wake of the 9/11 attacks the United States government actually asked American television networks to carefully screen videotaped communiqués issued by Osama Bin Laden, which it suspected might contain encrypted orders to al-Qaeda sleeper cells in the United States, before broadcasting them.[85]

In a similar vein, consider, finally, the following ad posted in July 1986 in the *Washington Times*: "DODGE—'71, DIPLOMAT, NEEDS ENGINE WORK, $1000. Phone (703) 451-9780 (CALL NEXT Mon., Wed., Fri. 1 p.m.)"[86] Although potentially available to the thousands of the newspaper's readers, only the KGB, who ran it, and FBI-agent-about-to-become-Soviet-spy Robert Hanssen, who had requested several weeks earlier that they run it, were actually aware that this seemingly innocuous "background noise" that resembled so many other "For Sale" ads was in fact a critical signal that indeed helped launch one of the most damaging espionage operations in American history.

CONTOUR DISTORTION

Setting things up to match (and thereby actually blend in with) their surroundings is not the only way to make them invisible, however. There is another major form of camouflage, for example, based on distorting, rather than simply blurring, a figure's contours (and thereby distinctive perceptual "signature"),[87] thus effectively breaking up its shape.[88] Such *contour-distorting camouflage* manifests itself throughout nature in the form of "boldly contrasting patches of color [that] mask the contour of their wearer, and 'break him up,'"[89] as so vividly exemplified, for instance, by the way many jungle birds' color patterns

> make their wearers invisible in the colored shimmering lights of southern forests. Certain parts of their bodies [thus] become visually cut out of the body and assigned to the background, so that the visible remainder takes on an irregular shape not characteristic of an animal.[90]

By essentially distorting the contours of the animal's silhouette, in other words, such patterns thus decrease its perceptibility.

By often involving perceptually complex figures, contour-distorting camouflage undermines our ability to actually spot objects. After all,

> it is the continuity of surface, bounded by a specific contour or outline, which chiefly enables the recognition of objects; so if the surface is covered with irregular patches of contrasted colours and tones, these patches tend to catch the eye of the observer and to draw attention away from the shape which bears them.[91]

Contrasting color patches that are placed at the edge of an animal's body, for example, thus tend to visually break up its outline.

A "busy" design, in short, thus compromises the recognizability of an object's shape, thereby threatening its perceived unity. By containing a lot of "internal detail that is more salient than [its] edge,"[92] an object may therefore no longer look like a single, perceptually coherent entity, and might-be perceivers may think that they are actually seeing several different objects.[93] Whereas by setting up a figure to match (and thereby blend in with) its surrounding background it might no longer be perceived as a *separate* entity, distorting its contours makes it more difficult to perceive it as a *single* entity.[94]

Thus, unlike drab uniforms, for example, contour-distorting military camouflage (see Plate 3)[95] is designed to actually distort,

FIGURE 3.1 Contour distortion helped make World War I ships less recognizable to enemy submarines. Courtesy Roy R. Behrens

rather than merely blur, soldiers' contours. It is likewise used to distort the outlines, and thereby break up the shape, of airplanes and buildings[96] in order to make them harder to target. In other words, it perceptually manipulates objects so

> that they will *no longer look like units*. . . . If a gun is covered with paint in such a way that one part of it will "fuse" with the bole of a tree, another with the leaves, a third with the ground, then the beholder will *no longer see a unit*, the gun.[97]

Indeed, in modern warfare, as so vividly exemplified by the "dazzle" paint schemes (see, for example, Figure 3.1)[98] used in World War I to make ships' shapes less recognizable to enemy submarines (and thus effectively the contour-distorting equivalents of the ancient Roman background-matching tactic of dyeing warships' sails in blue to match the color of the seawater),[99]

> it has become a real art to make objects such as guns, cars, boats, etc., disappear by painting upon [them] irregular designs, the parts of which are likely to form units with parts of their

environment. In such cases the objects themselves *no longer exist as visual entities*, and in their place appear meaningless patches which do not arouse the enemy's suspicion.[100]

Yet as anyone looking for something in a cluttered basement, pantry, or drawer (or on a "busy" carpet, for that matter) knows, it is not only their own perceptual complexity but also that of their surroundings that makes things less visible.[101] Spotting a target, after all, presupposes its relatively homogeneous surroundings,[102] and the more perceptually complex their surrounding background, therefore, the harder it is to spot figures embedded in it.[103] The fact that small tumors and microcalcifications are effectively camouflaged by the "busy" background of overlapping tissue structures in which they are perceptually embedded on X-rays, for example, makes it quite difficult for anyone but expert radiologists to detect them.[104]

As so spectacularly demonstrated in Annie Leibovitz's photograph of Keith Haring in Figure 3.2, contour-distorting camouflage is most effective when it is used in combination with background-matching camouflage and both the figure and its surrounding background are perceptually complex. That is why when zebras, for instance, stand (see, for example, Figure 3.3) or move in groups, it is much more difficult for predators to target any particular one as potential prey.[105]

Contour distortion is actually one of the hallmarks of Cubism, an art movement that, as exemplified by Pablo Picasso's "Portrait of Wilhelm Uhde" in Figure 3.4, indeed tried to visually obliterate figure-like objects through a deliberate

> breaking of contours . . . so that a form merges with the space about it or with other forms; planes or tones . . . bleed into other planes and tones; outlines . . . coincide with other outlines, then suddenly reappear in new relations . . . parts of objects shifted away, displaced . . . *until forms disappear* behind themselves.[106]

That explains, for example, Guirand de Scévola's decision to specifically recruit Cubist artists to his *Section de Camouflage* in

FIGURE 3.2 Combining background-matching with contour-distorting camouflage. 1986. © Annie LEIBOVITZ (CONTACT PRESS IMAGES)

FIGURE 3.3 When zebras stand in groups, it is harder for predators to target any particular one as potential prey. Herd of Zebras. © Theo Allofs/Corbis

FIGURE 3.4 Contour distortion is a hallmark of cubism. Picasso, Pablo (1881–1973) © ARS, NY. Portrait of Wilhelm Uhde. 1910. Coll. Joseph Pulitzer Jr., St. Louis, Missouri, U.S.A. Photo Credit: Bridgeman-Giraudon/Art Resource, NY

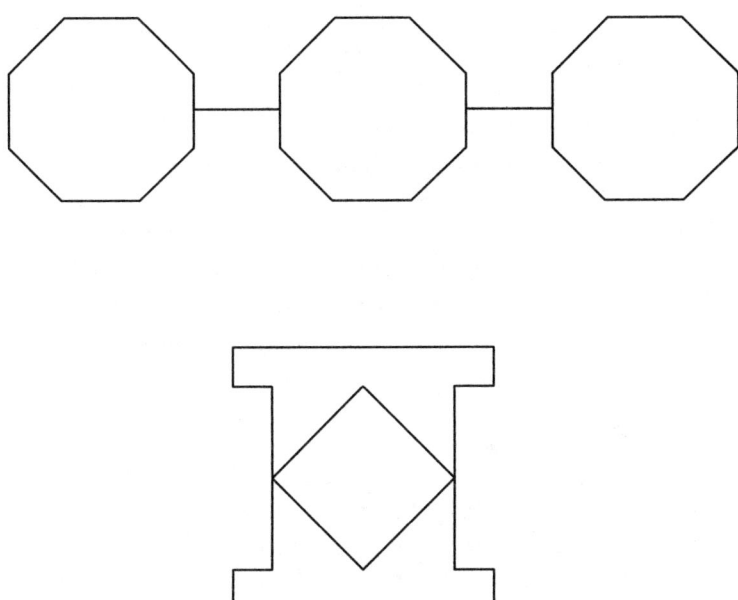

FIGURE 3.5 Even familiar shapes ("H," "K") are harder to spot when they are "hidden" in complex designs. Image courtesy of Ilanit Palmon

World War I. "In an effort to obliterate objects," as he later described it, "I used the techniques the Cubists had used . . . [T]his enabled me . . . to assign certain painters to camouflage who—because of their visual adeptness—could prevent the recognition of virtually any object."[107]

The cognitive challenge of spotting figure-like entities embedded in perceptually complex configurations has also inspired the Embedded Figures Test designed in 1926 by Kurt Gottschaldt to psychometrically measure our ability to mentally detach figures from their surrounding background and thereby spot familiar shapes "hidden" in complex designs[108] (such as the two Hs at the top, and the K at the bottom, of Figure 3.5)[109] or recognize a tune when presented with only three or four familiar notes acoustically embedded in a string of unfamiliar others.[110] It has likewise inspired treasure-hunt-like games such as

> the children's party game Hide the Thimble. An ordinary thimble is shown to all participants, and all but one leave the room, while the thimble is "hidden." . . . The thimble has to be hidden *in plain sight*. It may not be placed behind or under anything, or too high up for any of the children to see. . . . Once it is hidden, the rest of the children come back in the room and proceed to hunt for the thimble [and they] can usually be counted on to *look right at* the thimble several times without actually *seeing* it.[111]

Such playful use of contour-distorting camouflage is also exemplified by "Hidden Mickeys" in Disney theme parks;[112] word games (like Word Search and Boggle) in which players try to spot actual words "hidden" in a grid of scrambled letters; "hidden picture" puzzles such as the ones portrayed in Figure 3.6[113] and in the "Spot-It," I Spy, and Where's Waldo? children's book series;[114] and "knock-knock" jokes, the very essence of which involves distorting words' perceived auditory contours:

> Knock, Knock.
>   Who's there?
> Isadore.
>   Isadore who?
> Isadore [Is the door] locked? I can't get in.[115]

DIVERSION

There is an old Soviet joke about a factory guard who would occasionally see one of the workers walking out of the factory with a wheelbarrow full of trash. Suspecting that he might actually be trying to steal something, he would stop him and thoroughly search through the trash, yet he never found a single stolen item. Running into that worker years after they had both retired, he

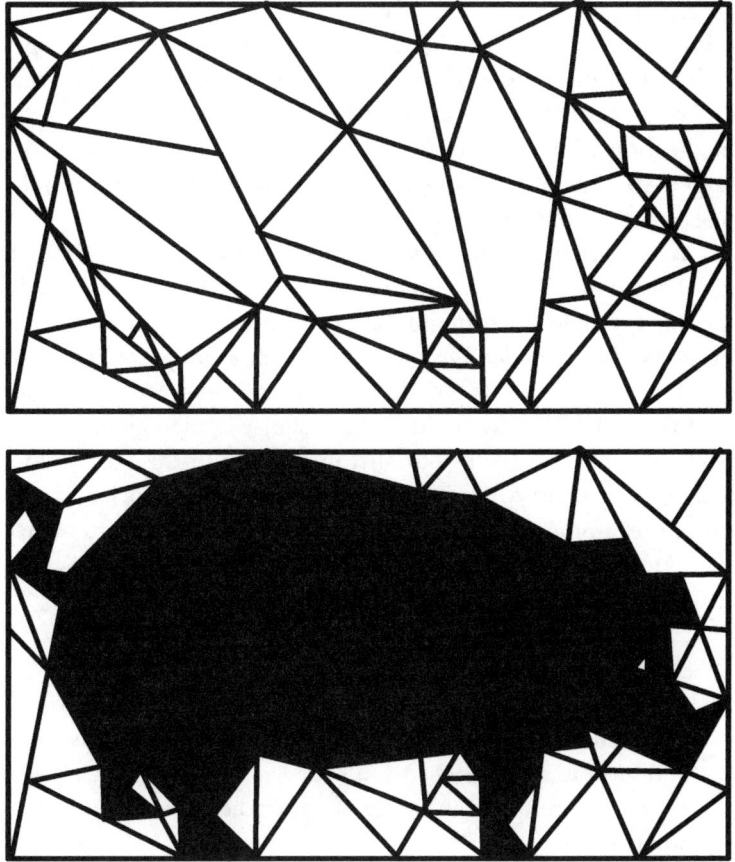

FIGURE 3.6 Can you spot the "hidden" pig in the top drawing? Courtesy Roy R. Behrens

could not resist asking him whether he had ever managed to steal anything from the factory. The man smiled and said: "Wheelbarrows." What by itself would have most likely constituted a highly noticeable figure was in fact "backgrounded" when filled with the trash, thereby becoming effectively invisible.

Unlike background matching and contour distortion, this form of camouflage involves deflecting attention from one potential target to another. After all, as we are so dramatically reminded by Simons and Chabris's "gorilla" experiment, it is especially when directing one's attention to some specific object that one often fails to notice others.[116] As when sugar-coating bitter medications, by diverting others' attention to other perceptual targets we thus often manage to prevent them from spotting the ones we wish to hide.[117]

As a result of such *diversion*, "signals" are sometimes mistaken for sheer "noise," and vice versa. Thus, as in background matching, for example, what might have otherwise been perceived as a figure is effectively *backgrounded*, as when a presidential limousine is strategically embedded in a motorcade of several identical-looking vehicles, or when we try to disguise "stolen glances"[118] as parts of general scans of our surroundings. On the other hand, it takes a full forty-five minutes into Alfred Hitchcock's *Psycho* (until the famous shower scene, that is) for the audience to realize that the film is actually not about embezzlement. As the master of suspense himself described this diversionary tactic,

> the first part of the story was a red herring. That was deliberate, you see, to *detract the viewer's attention* in order to heighten the murder. We purposely made that beginning on the long side, with the bit about the theft and her escape, in order to get the audience absorbed with the question of whether she would or would not be caught.[119]

Such diversionary tactics are quite common in magic shows. As Harry Houdini described it, conjuring is "the art of *making people look somewhere else*,"[120] thereby getting them to focus their attention on one point while "the action" is actually taking place at another.[121] Stage magicians thus try to create "areas of high interest" that capture the spectators' attention while the trick is in fact carried out in an "area of low interest,"[122] as when overtly waving one hand in the air while covertly pulling something out of their pocket with the other.[123] "Magic words" such as *abracadabra* are likewise designed to help them make the audience focus on a particular place and time while "cover[ing] up the fact that the trick actually takes place at a different place and a different time."[124]

Pickpockets often use similar tactics when faking distractive "accidents":

> When a pickpocket intends to rob you of your . . . wallet, he, or his confederate, takes care to distract your attention from what he intends to do by creating a diversion. He draws your eyes from what is really happening to what seems to be happening.[125]

He may thus "accidentally" bump into (or have his assistant "accidentally" spill coffee on) his victim, thereby diverting the latter's attention from realizing that he is at the same time sneaking the wallet out of his pocket.[126]

Fakes and feints (as in boxing, fencing, soccer, basketball, volleyball, ice hockey, and water polo, for example) operate on a similar principle. So, in fact, did Mao Zedong's Sun Tzu-inspired military tactic of "making a noise in the east while attacking in the west,"[127] a tactic also used in "Operation Fortitude," an elaborate deception campaign designed to conceal the Allied preparations to invade Europe through Normandy by diverting the Germans' attention elsewhere. The Allies thus faked, for example, an impending invasion of Norway from Scotland using various attention-grabbing diversions ranging from dummy landing barges and decoy tanks to false radio traffic and doctored reports from "German" agents.[128] They also pretended to prepare to land in Calais, and when attacking German coastal defenses in France sent out two bombing missions over the Pas de Calais for every mission over Normandy.[129]

Such diversionary tactics are sometimes also used by politicians to keep certain things out of the public's awareness.[130] They thus strategically time unpopular or embarrassing acts such as announcing controversial appointments or firing senior officials, for example, to coincide with other events that they hope will overshadow them. They likewise manufacture crises (and might even start wars) to divert the public's attention from economic problems or political scandals—a tactic also known since Barry Levinson's cinematic account of such an attempt as "wagging the dog." "To 'wag the dog,'" in other words, means "to purposely *divert attention from what would otherwise be of greater importance, to something else of lesser significance.*"[131] By doing so, of course, "the lesser-significant event is catapulted into the limelight, *drowning proper attention* to what was originally the more important issue,"[132] a perfect example of how something can actually be hidden in plain sight.

4

THE SOCIAL ORGANIZATION

OF ATTENTION

> The background does not consist only of what is empirically less salient or even less personally salient; it is also a reflection of what is less socially salient.
>
> ASIA FRIEDMAN[1]

NATURE AND CULTURE

Whether or not we notice something is partly a function of our physiology as human beings. Our perceptual access to the world, after all, is ultimately constrained by our senses, so that unlike bees, for example, we are unable to see objects reflecting only ultraviolet or shorter wavelengths of light, yet unlike rattlesnakes we also cannot see ones reflecting only infrared or longer wavelengths.[2] Our olfactory and auditory access to our surroundings is likewise rather limited compared to dogs and bats, for instance.

Such effectively species-specific physiological constraints are further exacerbated by the actual location of our sense organs. Unlike deer's, rabbits', and other prey animals' eyes, for example, which are essentially located on the sides of their heads, thereby allowing them a more panoramic view of their surroundings (so that no one can in fact sneak up on a rabbit from behind),[3] our own eyes are located on the front of our heads, thus allowing us a strictly frontal and therefore much narrower view.[4]

Given our physiological makeup as human beings, we are by and large very similar in what we are capable of noticing, yet we often differ from one another in what actually captures our attention.[5] While listening to a song, for example, some of us attend primarily to the lyrics, whereas others focus more on the music. By the same token, flipping through the same newspaper, people with different interests or concerns tend to notice different stories in it. And as sexist as that may sound to modern ears, when looking at the same woman a "breast man" may first notice her blouse while a "leg man" may first notice her skirt.

Yet not only the actual object of our attention but also our very "style" of attending often varies across individuals. After all,

> people differ in the extent to which they stress focusing in their cognitive organizations. They differ in their ability to maintain a . . . narrowed and selective attention that would permit them to screen out carefully from the stimulation which is relevant at any given time that which is irrelevant.[6]

While some of us, for example, are narrowly focused "screeners" who tend to selectively attend to only a few "relevant" stimuli and essentially ignore everything else, others are more widely focused "nonscreeners" who tend to also be aware of various other, "extraneous or irrelevant" stimuli.[7] In fact, we often differ from one another in our capacity to differentiate perceptual objects from their surroundings, as famously captured by the fundamental distinction introduced by Herman Witkin between "field-dependent" people who tend to perceive objects as effectively inseparable from their surroundings and "field-independent" ones who tend to be less influenced by context and to therefore perceive objects as more figure-like (that is, as discrete from their surrounding background).[8] The difference between these two *attentional styles* is manifested, for instance, in the relative ease at which we can identify "hidden" figures in visually complex designs such as in the Embedded Figures Test (or can spot Waldo, for that matter, in a crowded beach scene). Given the aforementioned considerable affinity between the ways we focus our perceptual and conceptual attention, it therefore also captures our varying propensities for thinking in a decontextualized, *analytical* manner.[9]

Yet the fact that we notice and ignore things not only as humans does not mean that we therefore also do so only as individuals. Indeed, although the way we differentiate figure-like, "relevant" objects from their background-like, "irrelevant" surroundings is by no means universal, nor is it entirely idiosyncratic. As a matter of fact, it is often very similar to the way some other people do so.

As exemplified by the distinctive manner in which individuals with autism, for example, fixate their gaze,[10] some such *non-universal yet also non-idiosyncratic* attentional patterns are biologically based, yet that is definitely not always the case. The remarkably similar manner in which structural engineers focus their attention when inspecting a building, for instance, is clearly not a function of their physiology. Nor is it any physiological constraint on our auditory capacity, for that matter, that keeps us from listening in on others' personal conversations.

That is indeed true of not only interpersonal attentional commonality but also intra-human attentional variability. Unlike the difference between eagles' and butterflies' respective ranges of vision, for instance, the difference between astrophysicists' and microbiologists' respective ranges of professional concern has nothing whatsoever to do with any physiological constraints on their ability to access the world through their sense organs.

Consider also, in this vein, the fact that men tend to filter "irrelevancies" out of their awareness far more readily than do women, as exemplified by their ability to attend to only one of several competing auditory messages, effectively ignoring the others.[11] Notice, too, the greater ease with which men manage to mentally detach figure-like objects from their background-like visual surroundings[12]—a tendency that has even led to the claim that their field of vision may actually be narrower than women's[13] as a result of the historical division of labor between male hunters and female child-rearers, the former developing an overall stance favoring focused attention and the latter developing one based on extreme vigilance and therefore favoring divided attention:

> As a nest defender, a woman has brain software that allows her to receive an arc of at least 45 degrees clear vision to each side of her head.... As a hunter, a man needed vision that would allow him to zero in on, and pursue, targets.... He evolved with vision almost like a racehorse with blinders so that he would not be distracted from targets, whereas a woman needed eyes to allow a wide arc of vision so that she could monitor any predators sneaking up on the nest.[14]

Yet even such seemingly natural gender-based *attentional division of labor*, after all, has always been as much about culture as it is about nature.[15] Indeed, in cultures that allow women to be more independent they also tend to be less "field dependent."[16]

By the same token, it is in fact as part of the unmistakably *socio*mental process of *moral attention*[17] that we actually come to regard objects lying beyond a certain socially delineated "circle of concern"[18] as morally irrelevant, thereby rarely extending our concern to also include, for example, the ants and cockroaches we so casually poison in our kitchens.[19] And it is likewise largely as part of the ultimately sociomental process of *erotic attention*[20] that we rarely come to experience our parents, siblings, and children as objects of our erotic desire.

In short, *we notice and ignore things not only as individuals and as human beings but also as social beings*. While it is certainly Nature that equips us with our sense organs, it is nevertheless our social environment that so often determines how we actually use them to access the world.

Effectively delineating the scope of our attention and concern, it is particular *attentional communities*[21] that often determine what we come to regard as relevant and to which we therefore attend. Such communities have their own distinctive *attentional traditions*[22] and therefore also distinctive *attentional habits* and biases, as manifested in their members' "tendency to notice . . .

specifically along certain directions."[23] It is specific conventions of what is noteworthy, for example, that make the Sistine Chapel and the Colosseum such "must" attractions for visitors to Rome,[24] and unmistakably social *attentional norms*[25] that likewise compel us to tactfully ignore another's stutter or open fly.

Such ultimately social traditions, conventions, biases, habits, and norms underlie our basic notions of relevance. As members of particular attentional communities we are thus "'perceptually readied' . . . to seek out and register those details that reflect our collective expectations, while overlooking other details that are equally perceptible,"[26] thereby being, for example, hyperattentive to the differences yet highly inattentive to the similarities between male and female bodies.[27] Such perceptual readiness manifests itself in the form of a mental filter that "let[s] in culturally meaningful details while sifting out the culturally irrelevant."[28] Selective attention is thus a result of being not only personally but also socially selective.

This definitely calls for a *sociology of attention*.[29] The voluminous amount of psychological and neuroscientific research on attention notwithstanding, only a pronouncedly sociological perspective can help reveal the unmistakably social attentional traditions, biases, habits, norms, and conventions underlying the way we access the world as members of particular attentional communities. Such a perspective, as we shall see, also helps highlight the political as well as the collective dimensions of human attention.

SOCIO-ATTENTIONAL PATTERNS[30]

The most visible manifestation of the social underpinnings of human attention are attentional patterns that are shared by some individuals (and therefore evidently not entirely idiosyncratic) yet nevertheless not by others (and thus also far from universal). As we shall see, although the way we focus our attention often resembles the way some others do, it is at the same time also quite different from the way still others focus theirs.

Unlike the aforementioned difference between humans' and rabbits' attentional patterns, however, such variability is not in what we are physiologically capable of noticing but in our overall attentional style, as so vividly illustrated by culture-specific styles of mental focusing, some classic examples of which are the highly contrasting attentional styles respectively predominant in hunting and farming societies. In marked contrast to farmers, hunters, after all, must be able to rapidly differentiate potential food sources from the immediate surroundings in which they are visually embedded.[31] (Such necessity may in fact be further compounded by the visual composition of their physical surrounds. Living in the Arctic, for example, one must "develop certain perceptual skills merely to survive . . . [One] must first of all in order to hunt effectively develop the ability to isolate slight variation in visual stimulation from a relatively featureless array.")[32] That certainly requires focused attention, and members of hunting societies indeed tend to be significantly less "field dependent" than members of farming societies.[33]

Ecology, however, does not explain why Germans, for example, tend to be better than Russians at spotting "hidden" figures in visually complex designs.[34] Nor does it explain why Guatemalans and Samoans have a tendency to attend to multiple objects, or follow several events, simultaneously rather than focus, as do Americans, on only one thing at a time.[35]

Americans' single-focused (and thus pronouncedly selective) attentional style is quite consistent, of course, with the general Western tendency to mentally detach things (and thereby envision them as distinct) from their surroundings. As quite revealingly implied in the Roman historian Pliny the Elder's account of the origin of painting (a story about a young Greek woman's attempt to capture her lover's image by "trac[ing] the profile of his face, as thrown upon the wall by the light of the lamp"),[36] Westerners have long cherished the contours they envision separating figure-like objects from their background-like surroundings. As so evocatively embodied in the very term Witkin chose to connote focused attention, "field independence,"[37] this remarkably distinctive feature of Western civilization is in fact a product of its emphasis on independence. Indeed, cultures that promote social interdependence tend to adopt a somewhat less-focused attentional style than those promoting social independence.[38] After all, "[i]f one perceives oneself as embedded within a larger context of which one is an interdependent part, it is likely that other objects . . . will be perceived in a similar way."[39] "Attention to the social field," in other words,

> also entails greater attention to the physical field. . . . Independent social systems foster analytic thought because the individual can focus on relevant objects without paying so much attention to the way they are enmeshed with other people and their goals.[40]

Spectacularly revealing in this regard is the striking contrast between Westerners' and East Asians' attentional styles. Since East Asian societies are generally much less individualistic than Western societies, claims Richard Nisbett, their members tend to view things as organically embedded in specific contexts.[41] Westerners, on the other hand, basically view the world as made up of discrete objects that are detachable from their contexts[42] and therefore tend to pay much less attention to contextual information:

> Westerners tend to engage in context-independent and analytic perceptual processes by focusing on a salient object independently of its context, whereas Asians tend to engage in context-dependent and holistic perceptual processes by attending to the relationship between the object and the context in which the object is located.[43]

Effectively ignoring contextual information,[44] Westerners can thus spot objects much more easily even when they are perceptually situated against an unfamiliar background.[45] East Asians, by contrast, notice much more contextual detail, yet have greater difficulty differentiating figures from their backgrounds.[46] They also tend to distribute their attention among multiple objects simultaneously rather than focus, as Westerners tend to do, on only one object at a time.[47]

The fundamental contrast between those two attentional styles is quite evident, for example, in studies tracking eye movements (those "unconscious adjustments to the demands of attention" which therefore reflect the way we actually allocate it)[48] as well as in brain scans monitoring neural responses to visual environments,[49] both of which show that Westerners tend to fixate sooner as well as longer than East Asians on foregrounded objects and are much less attentive to backgrounds.[50] It is also manifested in their highly contrasting aesthetic predispositions. Whereas Westerners exhibit a clear preference for paintings, drawings, and photographs that provide only a limited amount of contextual information,[51] East Asians tend to favor ones that offer viewers a great deal of context.[52]

What people tend to notice and ignore also varies across different *attentional subcultures* within the same society, however, as so vividly exemplified by profession-specific attentional traditions, conventions, and habits. Every profession, after all, has a distinctive sense of relevance[53] and therefore also distinctive concerns.[54] Depending on their profession, one person is thus likely to "notice details to which another person is blind ... [P]odiatrists notice feet, chiropractors notice posture and spinal alignment, orthodontists notice jaw alignment, dancers notice leg alignment,"[55] and so on.

Furthermore, professions often vary in the overall style of attending they implicitly and sometimes even explicitly promote. Consider, for example, the way surgeons, while operating, usually focus their attention only on a very small part of the patient's body selectively displayed through a hole in the surgical drape covering it, thereby mentally reducing his or her entire body to that "visually cut out" part,[56] that specific "piece of body below the operating lighting."[57] Needless to say, the contrast between such a pronouncedly *decontextualized* style of attending and the one prevalent among both landscape and anti-aircraft defense system designers, for instance, could hardly be starker. Nor, for that matter, could the somewhat analogous contrast between mathematicians'[58] and social workers' respective levels of context awareness. By the same token, whereas experimental researchers are specifically trained to focus their studies on a few preselected variables and essentially disregard everything else, both police detectives and investigative reporters are in fact trained to look for evidence "everywhere" and therefore to pay attention to practically "everything."

What we consider relevant and therefore tend to notice also varies across different settings and social occasions within the same society. While we are unlikely to notice whether other people around us happen to be chewing gum in an amusement park or at a picnic, for example, the very same act would most likely capture our attention at a formal reception or a funeral. By the same token, whereas players' gender is deemed irrelevant when playing dominoes or Parcheesi, it is considered highly relevant in basketball and soccer competitions, which are in fact even formally organized by participants' gender.[59]

Furthermore, the very decision whether something ought to be considered relevant and therefore attended to is often contested. Whereas some people, for instance, consider cyberbullying a serious problem, others do not think it even merits public attention. Disputes over whether applicants' race ought to play a major role, and therefore be explicitly attended to, in college admissions further exemplify such *attentional battles*. So do arguments over the relevance of politicians' sexual indiscretions and ethical debates over whether experimenting with animals warrants our moral concern.

What we tend to notice and ignore also changes with time,[60] as exemplified by historical shifts in what we deem noteworthy enough to feature in our museums. Thus, for example, we have recently witnessed

> a marked broadening of the objects deemed worthy of being preserved.... No longer are people only interested in seeing either great works of art or artefacts from very distant historical periods. People increasingly seem attracted by representations of the "ordinary," of modest houses and of mundane forms of work.[61]

By the same token, it is only relatively recently that food critics, for instance, have begun to include the noise level of venues in their restaurant reviews.[62] And whereas only a few decades ago smoking was still considered a mere "background" (and thus often unnoticed) activity, a sort of "side" occurrence, so to speak,[63] nowadays, as numerous smoking bans[64] seem to suggest, it has clearly become a highly marked and therefore noticeable figure-like activity.

Our moral horizons, too, keep shifting, as we keep incorporating new, traditionally ignored objects into our circle of concern. Only recently, after all, have the hundreds of thousands of Africans who die from starvation and disease every year begun to register on Americans' moral radar screens. By extending legal rights to prisoners, captured enemy soldiers, children, and even the unborn, many societies likewise grant moral relevance now to entire populations whose legal standing was not so long ago virtually unthinkable.[65]

Historical shifts in what we notice and ignore are also evident in science. In fact, as Thomas Kuhn so wisely noted, many scientific revolutions are but perceptual revolutions in which scientists actually come to "see new and different things when looking with familiar instruments in places they have looked before."[66] It was William Herschel's discovery of Uranus in 1781, for example, the first discovery of a hitherto unknown planet in several millennia, that made astronomers mentally ready to notice additional ones,[67] as evidenced by the exceptionally rapid successive discoveries of the hitherto undetected four largest asteroids by three different astronomers between 1801 and 1807. By the same token, not until the publication of Sigmund Freud's *The Psychopathology of Everyday Life* in 1901[68] had the everyday phenomenon presently known as a "Freudian slip" ever been systematically attended to, and not until the publication of Edward Hall's aptly titled *The Hidden Dimension* in 1966[69] had anyone given serious scholarly attention to the highly structured organization of interpersonal distance in human interaction. Indeed, as Thomas Laqueur has so spectacularly demonstrated, nor did European anatomists pay much scholarly attention even to the "obvious" anatomical differences between men and women until the late-eighteenth century, as gender was becoming increasingly significant politically.[70]

NORMS AND CONTROL

As one might expect, attentional communities do more than simply offer their members suggestions as to what they might find noteworthy. Indeed, they also generate various norms that actually tell them what they *should* attend to and what *ought* to remain in the background. Thus, for example, it is

> not simply that certain body parts are more available for us to inspect, and it is therefore those empirically salient details that we attend. Although some details may in fact be more visually salient, that alone cannot account for what we notice. At times, in fact, social norms of attention direct us to seek out and attend physical details that are far from obvious and to ignore those that are technically more salient.[71]

It is the normative (rather than, say, physical) salience of such details, in other words, that makes us pay more attention to them.

Attentional norms may be physically embodied in actual road signs explicitly cautioning us to watch for falling rocks, crossing deer, merging traffic, or slippery road conditions. Quite often, however, they are only implicit, as exemplified by the *norms of moral attention* that only tacitly inform us that weeds and bacteria are essentially morally irrelevant, or the equally tacit *norms of mnemonic attention*[72] that actually determine what we come to regard as historically insignificant and therefore not worth remembering.

As implied above, attentional norms also tell us what would be considered wrong to attend. As such, they implicitly embody certain notions of *attentional deviance*.[73] People who do not attend to their health in accordance with society's conventional norms, for example, are thus viewed as sociomentally deviant—"'irresponsible' if they [do so] less regularly than dictated by norms, 'hypochondriacs' if they do so too frequently."[74] Being a devout fan of a mere film extra (that is, someone who is conventionally supposed to be considered part of the background and therefore effectively ignored) is likewise the first indication viewers get that something must be seriously wrong with the character played by Arsinée Khanjian in the film *Speaking Parts*. She is most certainly an attentional deviant.

The kind of ignoring mandated by attentional norms often involves not just inattention but also some active *disattention*.[75] Indeed, not only does society affect what we habitually inattend, it also tells us what we ought to actively *dis*attend. Ignoring something, in other words, is often a result of more than simply failing to notice it.

Such attentional avoidance usually presupposes certain *norms of disattention*.[76] Thus, for example, effectively as a result of various taboos against looking,[77] we normally refrain from staring at people, thereby displaying "civil inattention"[78] and avoiding "uncivil attention."[79] We likewise expect doctors and patients to collaboratively play down, through their attentional behavior, the potential sexual undertones of gynecological examinations. Doctors are thus supposed to make it clear that their strictly clinical gaze[80] takes in only medically pertinent facts, while patients are expected to complement such efforts to de-eroticize the situation by "glanc[ing] upward, at the ceiling . . . eyes open, not dreamy" and "avoid[ing] looking into the doctor's eyes."[81]

Similar *attentional taboos*[82] are in fact formally prescribed by law. Thus, when screening job applicants, for instance, employers are nowadays often legally required to effectively ignore their age, sex, skin color, marital status, religion, sexual orientation, and political affiliation, and formally exclude those variables from their considerations to ensure equal employment opportunity. By the same token, under the exclusionary rule, unlawfully obtained evidence presented in court is considered inadmissible, and the judge can in fact instruct the jury to formally disregard it.

Attentional avoidance may also take the more subtle form of tact, however, as when one simply refrains from asking others

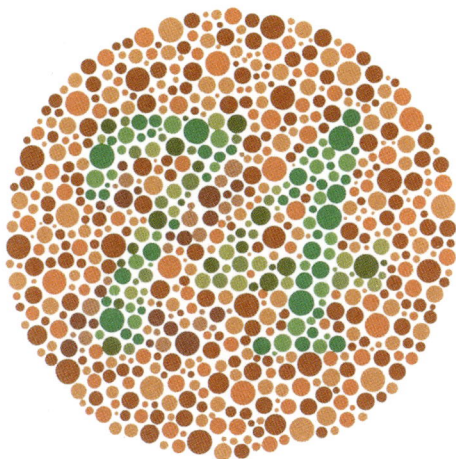

PLATE 1 "Color-blind" persons cannot differentiate the figure-like numeral "74" from the background in which it is visually embedded and are therefore unable to see it

PLATE 2 Blurring one's contours makes one "invisible." Liu Bolin, *Hiding in the City No. 96—Supermarket No. 3*, 2011. Photograph. Courtesy Klein Sun Gallery, NY. © 2011, Liu Bolin

PLATE 3 Without clear contours, a hunter becomes "invisible." Courtesy Realtree

PLATE 4 Even the conventionally backgrounded sky can be foregrounded as a figure-like object of attention. René Magritte, *Les muscles célèstes*. © Royal Museums of Fine Arts of Belgium, Brussels/Photo: J. Geleyns/Ro scan

PLATE 5 Look again at the "empty" space between the leaves. Sandro Del-Prete, *Secret Between Fall Leaves*. Copyright Sandro Del-Prete, www.sandrodelprete.com

PLATE 6 The figure-like silhouette of the skyline doubles as the background-like spaces between the shreds of a torn drape. Rob Gonsalves, *A Change of Scenery.* Courtesy of Huckleberry Fine Art

PLATE 7 Though featureless, the background is nevertheless an integral part of the flower arrangement. Copyright Peter Staes, www.imageguru.es

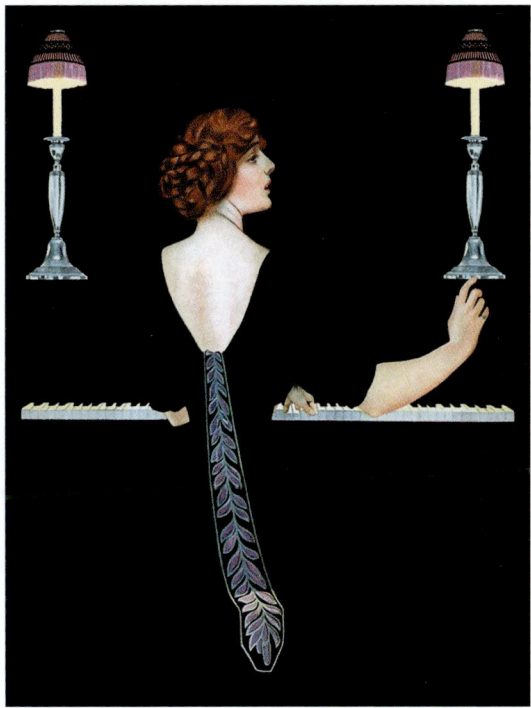

PLATE 8 Conventionally figure-like "objects" blending into their conventionally background-like "surroundings." Cole Phillips, *Good Housekeeping* October 1912 Cover

PLATE 9 Conventionally figure-like "objects" blending into their conventionally background-like "surroundings." Cole Phillips, *Between You, Me, and the Lamp Post*

PLATE 10 Because the woman's gaze is directed outside the frame of the painting, our attention is directed to the painting's conventionally irrelevant surroundings. Edgar Degas, *A Woman Seated beside a Vase of Flowers* (Madame Paul Valpinçon?) The Metropolitan Museum of Art. H. O. Havemeyer Collection. Bequest of Mrs. H. O. Havemeyer, 1929. Accession Number: 29.100.128

PLATE 11 This painting begs us to mentally "complete" the truncated geometrical forms beyond the canvas, thereby foregrounding its conventionally irrelevant surroundings. Piet Mondrian, *Tableau No. IV; Lozenge Composition with Red, Gray, Blue, Yellow, and Black*. C. 1924/1925. Image courtesy of the National Gallery of Art, Washington

PLATE 12 The sculpted figure literally flows out of the artistic frame into its supposedly irrelevant surroundings. Copyright Bill Mack

PLATE 13 A painted object literally steps out of the artistic frame into its supposedly irrelevant surroundings. Pere Borrell Del Caso, *Escaping Criticism*. Collection of the Bank of Spain

about "delicate" matters such as their marital problems or medical history. Indeed, as one can tell by how we view "nosy" people who do actually pry, tactlessness is clearly considered a major form of attentional deviance:

> Suppose you go around and find out how old everybody you meet is and how much they paid for their houses. Suppose each person with what you considered a physical oddity informed you in detail why he or she limped. . . . Suppose all single people explained to you why they were single . . . and every adult stated a rationale for the . . . nonexistence of his or her children. Suppose that upon greeting someone, you were able to find out immediately how old each piece of clothing he or she wore was, where it was bought, and for how much . . . [W]hy don't people quit asking such questions at every opportunity and go back to the system in which it was off-bounds to ferret information out of people and each person was allowed to volunteer topics he wished to discuss?[83]

Not only are we supposed to disattend certain things, we are also expected to at least feign inattention when we are unable to do so, such as by pretending not to have noticed one's facial birthmark or embarrassing faux pas.[84] In other words, aside from the pressure not to notice certain things, we are also socially pressured not to acknowledge the fact that sometimes we actually do notice them, as so poignantly captured in the tongue-in-cheek definition of a tactful man as someone who, having walked in on a naked woman, says: "Excuse me, sir."[85] As Goffman, who pioneered the study of such "tactful blindness,"[86] so aptly put it, "when bodies are naked, glances are clothed."[87]

As quite clear from the fact that they always entail a certain amount of pressure, norms are cultural manifestations of power and control, and power indeed plays a major role in what we come to notice and ignore. It is parents, after all, who decide what television shows their children can watch rather than the other way around. It is likewise the more powerful people in organizations who decide what gets to be included on staff meetings' agendas and also have the authority to "change the subject" (of attention) at those meetings.

Determining what others should consider relevant is a major form of *sociomental control*.[88] It is thus teachers, for example, who determine what students are required to read as well as what are the "main ideas" on which they ought to focus while reading and what are the mere "minor details" they should actually ignore as irrelevant.[89] In fact, curbing the scope of its citizens' attention is one of the hallmarks of the police state. During the 1930s and '40s, for example, people who happened to live near the Nazi concentration camps were actually instructed by the SS to turn their heads away, lower their gaze, and refrain from looking at the inmates or even at the trains transporting them to the camp.[90] And though they could certainly identify the unmistakable source of the stench coming out of the crematoria, they nevertheless learned to feign ignorance, trying to "look innocent by not noticing"[91] and pretending

> to ignore what they otherwise could not help but notice. [They] learned that if awareness of what was happening in and around the camp was unavoidable, one might still look away. Although cognizant of the terror in the camp, they learned to walk a narrow line between unavoidable awareness and prudent disregard.[92]

In so doing, they thus came to embody the kind of citizen who in fact makes the police state possible—"not looking, not even asking . . . not once curious."[93]

Although curbing the scope of others' attention may sometimes entail explicit acts of censorship such as closing down newspapers and radio stations and imposing Internet blackouts (that is, the macrosocial equivalents of "don't look" or "don't listen" taboos), there are also various tacit manifestations of the power to control what others attend to, as exemplified by the lists of "must-see" destinations on tourist itineraries[94] or even by the way seats are traditionally positioned in classrooms.[95] Consider also in this regard the tacit attentional norms that pressure academic scholars to consider certain works canonical "must-reads" (and to thereby demonstrate their awareness of "the literature" by also regarding them as "must-cites"), implicitly ignoring others. They likewise often pressure them to confine their scholarly attention to specific "fields" and to even more narrowly delineated "areas" within them. Such tacit *intellectual blinders* certainly curb the scope of academic thoroughbreds' professional attention, which is why very few scholars are actually willing to endure the difficulties (such as in getting hired, funded, published, and promoted) often faced by attentional deviants who venture beyond the confines of the inherently parochial "attentional ghettos" of their academic discipline or choose to work on topics conventionally envisioned as lying on the margins of its awareness.

ATTENTIONAL SOCIALIZATION

As implied by the very notion of attentional norms, separating the relevant from the irrelevant is a sociomental act performed by members of particular attentional communities who *learn* to focus their attention on certain parts of their phenomenal world while systematically inattending or even disattending others in accordance with their community's distinctive attentional tradition, conventions, biases, and habits. As members of such communities, we thus essentially learn what to notice and what to ignore as part of our *attentional socialization*.[96]

It is hardly a coincidence, therefore, that the very first person in Hans Christian Andersen's classic story "The Emperor's New Clothes" to point out that the emperor has actually no clothes is indeed a presocialized child who has yet to learn what he is supposed to tactfully disattend[97] (such as by pretending not to hear or smell anything when someone around him breaks wind, for instance). By the same token, when they go to the zoo, young children have yet to learn

> to attend to . . . the cues that direct viewing [there]. While more mature visitors know that exhibits are contained within the zoo's

enclosures, and look there for "viewable" contents, very young children may ... ask about guardrails ... maintenance areas and equipment. ... *Learning to look properly* ... means coming to understand that some features deserve attention, while others are ... meaningless. (The child learns, for example, to see the animal, but not the cage.)[98]

Thus, when young visitors attend to conventionally "wrong" objects of attention such as the pigeons and squirrels that linger in between the zoo's enclosures, for instance, their parents indeed often try to redirect their attention to the specifically exhibited (and therefore conventionally "noteworthy") animals.[99] By the same token, when a young attentional deviant points to an elephant statue, a corrective response such as "I see that statue—but look at the *real* animal"[100] is thus clearly part of his attentional socialization.

The psychological as well as biological foundations of human attention notwithstanding, it is nevertheless the attentional traditions, conventions, norms, and habits we internalize during our attentional socialization that often determine what we come to notice and ignore.[101] We thus learn, for example, that whereas its author's last name is particularly relevant for finding a book in a bookstore, the color of its jacket is not, as well as how to curb our moral concerns in a socially appropriate manner. It is not just our personal feelings, therefore, that make us concerned about some war casualties (women, enemy civilians) more than others (men, enemy troops) and deem mosquitoes morally irrelevant.

Attentional socialization can be quite explicit, as in special "appreciation" courses where we are effectively taught by expert *attentional mentors* what we should specifically attend to when looking at a painting, listening to a piece of music, watching a film, or tasting a glass of wine, or as children watch their reading teacher demonstrate how to identify the text's "main idea" and separate "key" (that is, relevant) information from "minor" (that is, irrelevant) details.[102] Most of it is implicit, however, as we usually learn what to attend to and what to ignore by simply noting what others around us attend to and ignore. By watching his mother describing their trip downtown that morning, for example, a young child thus also gets a tacit lesson in what is considered attention-worthy and what is conventionally deemed mere "background noise." By the same token, by watching televised broadcasts of football games and track meets, he tacitly learns what spectators are actually supposed to attend to in such perceptually complex situations, in much the same way as new members of an organization tacitly learn from reading the official minutes of the first few meetings they attend what is considered noteworthy at such events.

Such tacit attentional socialization is also part of the process of learning a language. In fact, even what we actually see is partly "a function of . . . the words that are available to name visible things" and thus constitute "programmes for perception":[103]

> This point was recently reinforced for me as I settled into a new job and became acquainted with a new group of colleagues. Long before I had learned everyone's names and roles, I knew with certainty whether they were male or female, whereas it took me over a week to register that one of the people I was working most closely with has one brown eye and one blue eye. Of course, if language were organized around eye color rather than sex, I would have been unable to ignore it.[104]

By the same token, native speakers of a noun-dominated language such as English[105] are tacitly socialized to view animals as members of specific taxonomic categories, quite independently of the particular contexts in which they are situated. (That may also be a consequence of visiting the Reptile House or Monkey House at the zoo, in marked contrast to the way in which watching the television series "Planet Earth," for example, may actually lead one to view them as integral parts of the multi-species ecological communities in which they are organically embedded.)[106]

Membership in a given attentional community is often characterized by the ability to notice things that nonmembers tend to ignore as well as to ignore some of the things they attend to, and that is particularly evident in the case of professions, which are indeed often characterized by their distinctive foci of attention and concern. One of the most important aspects of becoming professionally socialized, in fact, is acquiring the ability to notice the things that are considered relevant[107] and ignore those that are deemed irrelevant by one's profession, and attentional socialization is therefore a critical part of the process of professionalization.

Attentional patterns, as we already saw earlier, often vary from one profession to another, and while an orthopedist, for example, is unlikely to ask patients with shoulder tendinitis about their appetite, which he considers totally irrelevant, a holistic practitioner might very well do so, having been professionally socialized to regard every part of the body as organically related to all the others. That does not necessarily mean, however, that holistic practitioners are therefore more curious individuals than orthopedists. Nor, for that matter, is it their superior sensory capacity that enables professional chess players to visually scan a larger area of the chessboard than laypersons[108] and professional soccer defenders to be aware of more potential attackers than just the one currently possessing the ball.[109] And just as it is not simply their temperament that makes flight attendants seem friendly and bill collectors seem hostile,[110] it is also not necessarily customs officers' and security personnel's paranoid tendencies that make them particularly vigilant or division commanders' wider scope of personal concern that enables them to think more strategically than tactically oriented platoon leaders. Such seemingly personal "curiosity," "superior sensory capacity," "paranoia," and "concern" are after all but products of the particular manner in which the individuals occupying those professional positions are attentionally socialized.

Such socialization is indeed a big part of what distinguishes experts (such as the sports commentators we listen to while watching televised gymnastics and diving competitions)[111] from novices. Whereas untrained viewers of a painting, for example, usually notice only foregrounded objects in it, art-trained viewers also attend to its backgrounded features.[112] By the same token, when sight-reading, trained musicians are much better than untrained ones at noticing sudden changes of tempo, meter, or key signature. And when listening to polyphonic music, they do not just "focally process the upper part alone" while "relegating the lower material to a background role."[113] A greater use of one's peripheral vision (and thus a wider range of horizontal scanning) likewise distinguishes experienced from novice drivers.[114]

Similarly, when viewing a chest X-ray, for instance, beginning radiologists tend to spend a lot of time examining its peripheral portions, effectively looking for radiographic findings outside the lungs, pleura, and mediastinum, which expert radiologists have already learned from years of experience are often but background noise.[115] That implies, of course, the latter's ability to distinguish "relevant" from "irrelevant" information, in marked contrast to the former, who "may not recognize relevance and irrelevance" and therefore "attribut[e] prominence to items of little import" while failing to notice the most critical findings.[116]

It is partly through learning their profession's specialized vocabulary that members come to acquire its distinctive attentional habits. In acquiring such a vocabulary, a new member is actually acquiring "a set of colored spectacles. He sees a world of objects that are technically tinted and patternized. A specialized language constitutes a veritable a priori form of perception and cognition."[117] Yet as exemplified by the way we sort out the injured in battlefields, disaster areas, and hospitals based on their need for urgent treatment, professional socialization is partly also sensorial. As a process of prioritizing attention,[118] triage requires the ability to spot critical figure-like signals in perceptually complex, "noisy" environments, and emergency-room receptionists and triage nurses are indeed specifically trained to spot visual (massive bleeding), auditory (shortness of breath), and tactile (very high fever) symptoms that require immediate medical attention.[119]

Occasionally, professional socialization may actually involve reversing ordinary attentional habits, as in the case of interpreters of aerial stereoscopic photographs, the foreground and background planes of which are both simultaneously in focus,[120] or for that matter soccer forwards, who must learn to spot traditionally backgrounded "negative" spaces such as the supposedly empty gaps between opponent defenders or between goalkeepers and goalposts.[121] A somewhat similar cognitive challenge is also faced by psychotherapists, who are professionally socialized to attend not only to what their patients are saying but also to what they are *not* saying.

The considerable extent to which a profession's distinctive attentional tradition and habits affect what its members come to notice is also evident in science. After all, what scientists actually "observe" is partly a product of the particular way they focus their attention as a result of their professional socialization.[122] "Only after . . . such transformatio[n] of vision," indeed, "does the student become an inhabitant of the scientist's world, seeing what the scientist sees."[123] Only their professionally acquired sociological imagination, for instance, enables sociologists to envision social movements, labor markets, power structures, influence networks, and kinship ties.[124]

Furthermore, when doing research, scientists confine their scholarly attention to a few selected variables, thereby implicitly inattending, or even explicitly disattending, to others.[125] A decision to examine the relation between offenders' age and the amount of time it takes them to become eligible for parole, for example, thus also implies a tacit prior decision on the researcher's part to essentially disregard their reading habits and table manners. A decision to examine the relation between having children and the amount of time it takes students to graduate likewise implies a tacit prior decision to effectively ignore their height and cholesterol level, and the researcher would most likely not even try to control for such presumably irrelevant variables (in marked contrast to her explicit effort to disattend to, so as to control for, their grade point average, for instance).

Some forty-plus years ago I was once invited to visit Robert Freed Bales's seminar on small-group behavior and, along with his students, watch a group of people interacting in his lab. Later, as we compared the notes we took while observing them, I noticed that whereas most of the other students' notes were about the power dynamics within the group, mine were about eye contact, impression management, and use of space. Such fundamental difference in focus, however, had less to do with our personal observational skills and more to do with the fact that, at the time, I was studying with and greatly influenced by Goffman, whose notion of social interaction was significantly different from that of Bales. Having been professionally socialized into two very different attentional traditions, it was as if we were actually observing two different groups.

COLLECTIVE ATTENTION

A sociology of attention also helps remind us that *we actually notice and ignore things not only as individuals but also as collectives*.[126] After all, there are many situations—from kissing, ballroom dancing, playing ping-pong, or having a conversation to televised presidential debates, Oscar ceremonies, New Year's Eve ball drops, and state funerals—where two or more people jointly sustain a single focus of attention simultaneously.[127] Indeed, as so spectacularly demonstrated by the hundreds of millions of people around the world who effectively watch the televised opening ceremony of the Olympic Games or the soccer World Cup final "together," Marshall McLuhan's once-futuristic "global village" is no longer just a metaphor.

Attentional "togetherness" may not always imply simultaneity, yet it does imply a *collective focus of attention*. In other words, it presupposes a *shared sense of relevance*[128] and therefore implicitly also a shared notion of irrelevance.

That implies, of course, some basic agreement about what we should attend, inattend, and disattend to. At the microsocial level of a conversation, for example, it is usually manifested in the form of a tacit agreement to focus our attention on what is being said, thereby effectively ignoring presumably irrelevant "background noise" such as scratching and postural adjustments.[129] By the same token, at the macrosocial level, it typically assumes the form of a common ranking of issues in terms of their collectively perceived relative salience. While some of them become part of the so-called *public agenda*, others remain off our collective radar screen.[130] We thus focus our collective attention on abortion, gas prices, and immigration, for example, while collectively ignoring poor service, dangerous driving, and elderly abuse.

Viewing "social problems" simply as reflections of objectively problematic situations cannot explain, therefore,

> why some conditions are defined as problems, commanding a great deal of societal attention, whereas others, equally harmful or dangerous, are not.... Why are conditions and events in the Third World that affect the life chances of millions of people ... the object of only the most cursory and superficial public attention ... ? Why do toxic chemical wastes in landfills receive more public discussion than dangerous chemicals in America's work-places?[131]

To answer such questions, one needs to examine how we collectively come to attend to certain issues more than others. As the process of "placing an issue . . . on the public agenda so that it becomes the focus of public attention,"[132] *agenda setting* is thus the group-level analogue of the mental process through which attention is selectively allocated at the individual level.[133]

Nowadays, of course, it is by and large the mass media— newspapers, radio, television ("the most pervasive and efficient system for the management of attention"),[134] and the Internet— that actually set the public agenda.[135] By figuratively *spotlighting* certain issues and events while downplaying or even completely ignoring others, they in effect shape our collective sense of relevance, thereby basically determining what we collectively attend and inattend to.

Thus, by choosing which issues and events make newspaper headlines and become the lead stories on radio and television newscasts, for example, editors and news directors actually affect what we come to consider the most important issues and events.[136]

"The lead story on page 1, front page versus inside page, the size of the headline, and even the length of a story all communicate the salience of topics,"[137] thereby determining their perceived public relevance. Indeed, as Bernard Cohen so aptly put it, though they may not always succeed in telling us what to think, the media are therefore nevertheless "stunningly successful in telling [us] what to think *about*."[138]

In fact, they also determine the length of our *collective attention span*. As soon as they stop "covering" (and thus effectively directing our attention to) a particular news story we have been following for several days or even weeks, we often forget about it, thereby exemplifying the operation of an unmistakably media-driven *public attention cycle* whereby issues and events gain public prominence, remain on our collective radar screen for some time, and then gradually fade from public awareness once media coverage ends.[139] Even the most sensational so-called *focusing events*[140] (hurricanes, mass shootings, political scandals), after all, actually have a relatively short public attentional "shelf life," inevitably receding to less prominent spots on our collective radar screen after a while and ultimately dropping off it altogether.

Furthermore, the media also keeps certain things off the public's radar screen simply by not covering them. That, in fact, happens not only in totalitarian states, where all the newspapers, radio stations, and television networks are effectively controlled by the government, but even in more pluralistic political systems, as exemplified by the publicly invisible "homecomings" of America's fallen soldiers during the Iraq War, the considerable difference in media coverage when white and black American children go missing,[141] or the general reluctance on the part of the American press to cover "minor" candidates it believes (and in a self-fulfilling manner thereby also helps ensure) are not going to play a major role in an upcoming election.[142]

In fact, even in telling us what to actually think about, the media are implicitly also telling us what *not* to think about. As such, they always play a critical role in both the production and maintenance of our *collective blind spots*.

5

CONCLUSION

> For by narrowing my attention . . . I no doubt increased my capacity to understand the small part I was actually looking at, but I also increased that dark area outside its concentrated beams in which ideas might remain unrecognized.
>
> MARION MILNER[1]

A testament to the importance of selective attention in our lives, "[p]art of normal human development is learning to notice less than we are able to."[2] For one thing, such "reduced awareness"[3] helps protect us from the predicament of constant mental bombardment. After all, at any given moment,

> thousands of stimuli are bombarding [our] sense organs . . . [W]e can see and hear a vast number of things [but] can at any one time [attend] to only a small number of them. Stimuli will thus be competing for [our attention] and there must be ways of determining which stimuli will win the contest.[4]

> To a large extent adjustment to the environment . . . develops by a process of selection and inhibition of incoming sensory input, so that only part of [it] is effectively registered in consciousness.[5]

That presupposes, of course, the ability to actually "tune out"[6] irrelevancies, and indeed, one of the functions of our brain is

> to protect us from being overwhelmed [by] irrelevant knowledge, by shutting out most of what we should otherwise perceive . . .

and leaving only that very small . . . selection which is likely to be practically useful.[7]

The organism is literally bombarded by sensory stimuli . . . which are too numerous to process completely. However, the human brain has been optimized in the course of evolution to select salient information by directing the focus of attention, which permits the *filtering* of information relevant for everyday life.[8]

To appreciate this screening, or filtering, function of our brain,[9] consider people with schizophrenia, whose selective attention is effectively "reduced to zero," according to the person who coined that term, "so that almost everything is recorded that reaches the[ir] senses."[10] As such, they are practically flooded with incoming sensory data,[11] which is indeed the way they themselves describe the "over-inclusive" awareness[12] that precludes them from attending to anything in a selective, "focused" manner:

> I do want to explain . . . the *exaggerated state of awareness* in which I lived. . . . Each of us is capable of coping with a large number of stimuli, invading our being through any one of the senses. . . . It's obvious that we would be incapable of carrying on any of our daily activities if even one-hundredth of all these available stimuli invaded us at once. So the mind must have a filter . . . sorting stimuli and allowing only those which are relevant to the situation in hand to disturb consciousness. . . . What had happened to me . . . was *a breakdown in the filter,* and a hodge-podge of unrelated stimuli were distracting me from things which should have had my undivided attention. . . . I had *very little ability to sort the relevant from the irrelevant. The filter had broken down.*[13]

Effectively "bombarded with irrelevant information," people with schizophrenia thus feel "swamped by the incoming tide of impressions,"[14] to the point where they may actually close their eyes and stuff their ears with cotton wool.[15]

Furthermore, as anyone who has ever literally or figuratively "looked for" an object or an idea knows, it is our ability to "[s]elec[t] what to attend to and maintai[n] *concentration* on that thing, rather than on irrelevancies"[16] that actually allows us to perceive or conceive any "thing" at all.[17] Being able to mentally detach figure-like objects from their background-like surroundings is an absolute prerequisite for finding a paper clip in a cluttered kitchen drawer as well as for conducting a successful air search for shipwreck survivors or detecting a lung tumor on an X-ray. Such ability to "'zoom in' on task-relevant information while ignoring distractions"[18] is also a major professional requirement for surgeons and jewelers, pole-vaulters and simultaneous interpreters, proofreaders and auctioneers.

Yet our ability to attend to the world in a selective manner inevitably also blinds us to everything else that might have otherwise "entered" our mind. By fixating on his direct opponent, for instance, a soccer player may thus fail to notice, and thereby pass the ball to, an open teammate on his right.[19] Focusing too heavily on one possible solution may likewise impede one's ability to consider others.

This clearly poses a problem, particularly given the inverse relation between perceptual salience and empirical prominence. After all, as we saw earlier, the marked parts of our phenomenal world, which we explicitly notice, are proportionally smaller than the unmarked ones, which we implicitly ignore. In other words, only a small part of what is phenomenally available to us actually becomes the focus of our attention. *There is much more we could potentially perceive or conceive had we not deemed it irrelevant.*

Indeed, selective attention involves a considerable element of *mental constriction*,[20] thus effectively "closing" our minds. As William Blake so evocatively put it, "[i]f the doors of perception were cleansed every thing would appear to man as it is, Infinite. For man has closed himself up, till he sees all things thro' narrow chinks of his cavern."[21] And "[b]y narrowing my attention," adds Marion Milner, while "I no doubt increased my capacity to understand the small part I was actually looking at . . . I also increased that dark area outside its concentrated beams in which ideas might remain unrecognized."[22] Such a restricted and therefore impoverished experience of reality is perhaps best captured by the metaphor of a visually restrictive tunnel. Essentially evoking the perceptual experience of seeing the world as if through a straw,[23] *tunnel vision* implies a highly parochial outlook and is thus effectively synonymous with *narrow-mindedness*.

Focusing on any "thing" and thereby differentiating it from its perceptual or conceptual surroundings also presupposes a certain degree of obliviousness to context.[24] As such, it inevitably also implies a strong element of denial,[25] since no amount of ignoring can really make what we disregard as irrelevant disappear. In other words, we cannot simply wish away what we figuratively banish beyond the narrow tunnels through which we access our phenomenal world. By keeping various sensations, thoughts, and feelings out of our awareness, such denial saps our energy,[26] ultimately numbing us intellectually as well as emotionally.[27]

MULTIFOCAL ATTENTION

But selective attention, distinctively associated with its single-focused character, is by no means our only mode of awareness. Just as common, for instance, is *multifocal*[28] (or "divided," "distributed")[29] attention, which, as we saw earlier in the case of Guatemalans and Samoans, for example, involves focusing on more than one thing at a time.[30] As evidenced by the practice of multitasking[31] (such as monitoring the road ahead as well as listening to the radio or talking with someone while driving),[32] our ability to *simultaneously attend to multiple foci of attention* certainly defies the notion that in order to attend to any given "thing" we must inattend or disattend to any others. As such, it represents a clear rejection of the rigid "either/or" logic[33] underlying our experience of figures and backgrounds.

Consider, for example, "The Word," a popular segment of the television show *The Colbert Report* in which political satirist Stephen Colbert delivers a serious-sounding monologue while at the same time flashing on the screen a facetious commentary on what he is saying. Although technically speaking those two seemingly competing targets of the viewer's attention may in fact be attended to sequentially, thereby involving rapid switching from one to the other, they are nevertheless phenomenologically experienced simultaneously.[34] Only by concurrently attending to both his spoken and written messages, indeed, can Colbert's viewers fully enjoy the comic effect of such farcical juxtapositions.

By the same token, consider also Cubist paintings (such as the one featured in Figure 3.4), those "infractions of conventional spatial syntax"[35] where the viewer is expected to *attend to several competing figure-ground configurations simultaneously*. Even if he actually switches rapidly from one such configuration to another, they are nevertheless phenomenologically experienced simultaneously. By featuring such visually dazzling compositions, Cubism clearly challenges our conventional view of the contours supposedly delineating figure-like objects of attention.[36]

In presupposing the viewer's multifocal attention, Cubist art also resembles polyphonic music, in which two or more voices are designed to simultaneously attract the listener's attention with equal force. (It likewise resembles polyrhythmic music, where listeners are expected to simultaneously attend to several competing rhythms.) Unlike in homophonic music, where one voice is perceived as a figure-like melody and all the others as its background-like accompaniment, in polyphonic music all voices are designed to be perceived as figure-like melodic lines of equal status.[37]

As such, polyphonic music implies a listener who can simultaneously attend to several separate strands of melody[38] (unlike the novice listener, who can do so only in rapid succession). In other words, as Anton Ehrenzweig points out, it actually requires a special kind of listening, because while

> the single-melody-minded listener ... still follows that voice in which he has just picked out the theme, he will often discover that yet another voice has already taken up the theme a few

bars before and [that] he ought to have switched over his attention much earlier. . . . In order to enjoy polyphonic music a change of attitude is necessary. One has . . . to follow the unfolding of this structure with a *diffuse attention* not concentrated on a single voice but on the structure as a whole . . . [O]nly then will the listener feel the deep elation connected with polyphonic music which has to speak in several tongues instead of in one.[39]

Only then, indeed, can one fully appreciate the intricate sonic texture jointly produced by the flute and the bassoon in the Aria movement from Heitor Villa-Lobos's *Bachiana Brasileira No. 6*, let alone the exquisite beauty of the vocal and instrumental "multilogues" in the Sanctus movement of Giovanni Palestrina's *Missa Papae Marcelli* and the spectacular six-voice (!) ricercar from Johann Sebastian Bach's *Musical Offering*.

OPEN AWARENESS

Yet the challenge to the way selective attention inevitably narrows[40] our field of consciousness comes not only from multifocal attention. After all, phenomenologically speaking, despite the fact that the background-like parts of our phenomenal world are clearly "less intensely grasped," they are nevertheless "dimly and *marginally* present."[41] In other words, as implied by the use of the margin metaphor here,[42] we do also attend to the "periphery" of our awareness.[43]

The implicit analogy to our *peripheral vision*[44] is almost self-evident. Our eyes, after all, contain not only cone-like photoreceptor cells ("cones") that enable us to perceive figure-like objects but also rod-like ones ("rods") that allow us access to their background-like surroundings.[45] And while our central (foveal) vision is indeed narrowly focused, our peripheral vision offers us a wider, more panoramic[46] view of our surroundings,[47] as evidenced, for example, in driving, juggling, boxing, soccer, handball, volleyball, and basketball,[48] where seeing "the whole court" is in fact considered a form of art:

> Most people remember the incredible passing abilities of Earvin ("Magic") Johnson, who became famous for his "no-look passes." Magic Johnson [used] to fool his opponents by looking in the direction of the most obviously open teammate, but then passing the ball to another player instead without even looking at him.[49]

As one might expect, such a wider scope of attention tends to promote a more "open" mode of awareness[50] that sharply contrasts with narrowly focused selective attention:

> With *focused consciousness*, a person is totally on a task, absolutely absorbed in it. The ticking of the clock . . . the sounds from the street . . . all vanish into a gray background, as the bright light of consciousness illuminates only the single task. *Open consciousness*, on the other hand, is diffused . . . [It] touches the ticking of the clock [which is then] interrupted by the sound of a truck rumbling by on the street outside.[51]

A perfect example of such mental openness is *mindfulness*. In marked contrast to concentration, a pronouncedly exclusionary mode of awareness in which we essentially ignore everything other than the object on which we focus our attention, it is an all-inclusive state of mind[52] where we actually try to be aware of *everything* we experience, thereby rejecting the very notion of selective attention. (Unlike concentrative meditation, for instance, in which meditators narrow their attention to a single source of stimulation thereby withdrawing their awareness from anything that might distract them from it, mindfulness-based Vipassana meditation involves expanding one's attentional scope[53] thereby promoting "an open awareness to *all* stimuli in an undiscriminating fashion.")[54] Rather than focusing on a single object, it thus promotes "a field of experience [that] does not have a finite horizon."[55] As such, it is "awareness of *all* stimulation,"[56] including thoughts and feelings we usually suppress as distractions and thereby avoid experiencing.[57]

In marked contrast to being attentionally selective, which inevitably implies an element of mental constriction, widening the scope of one's attention (even literally, in the form of a wide-eyed look)[58] enhances creativity[59] (which may indeed explain why highly creative individuals are often also more distractible).[60] Being creative, in other words, presupposes a certain degree of mental (perceptual as well as conceptual)[61] openness:

> The highly creative individual may be privileged to access a greater inventory of unfiltered stimuli . . . thereby increasing the odds of original recombinant ideation. Thus, a deficit that is generally associated with pathology may well impart a creative advantage.[62]

Indeed, one of the most distinctive characteristics of being creative is the ability to notice what others tend to ignore,[63] which requires widening the scope of one's attention as well as being mentally ready *"not to screen out the irrelevant."*[64] Such "readiness to assimilate a wider range of available input" as well as "to *regard all information as potentially relevant*"[65] comes particularly handy, of course, in situations requiring original, "out-of-the-box" thinking.

RELEVANCE RECONSIDERED

Not only is it important that we examine other modes of awareness besides selective attention, we also have to be careful not to essentialize the seemingly intrinsic figure-like or background-like character of what we conventionally consider "figure" or "background." As anyone who has ever wondered whether zebras are white animals with black stripes or black animals with white stripes knows, *nothing is inherently figure-like or background-like.* Figureness and backgroundness are not inherent qualities, and as exemplified by the two yin/yang-like faces in Figure 5.1,[66] any component of our phenomenal world can in fact be perceived as *both figure and background.*[67]

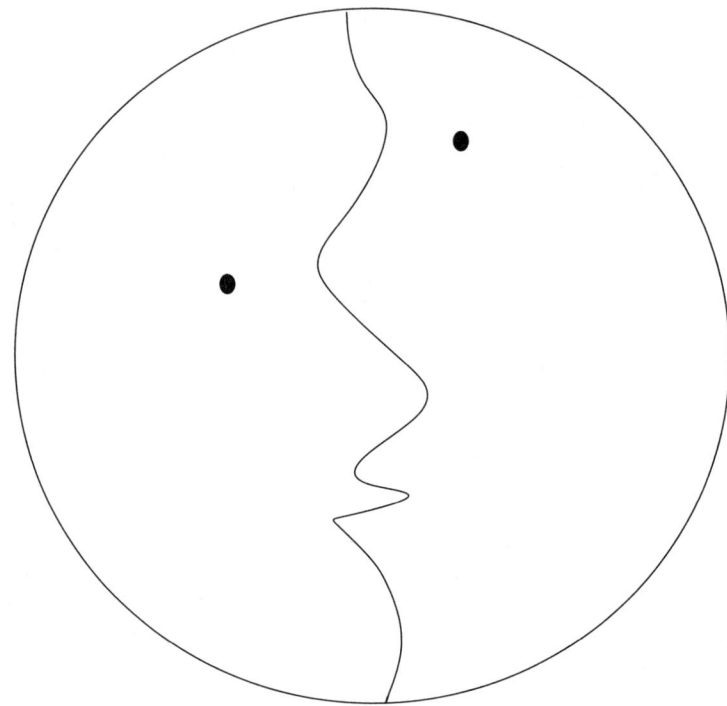

FIGURE 5.1 Nothing is inherently figure-like or background-like. Image courtesy of Ilanit Palmon

Attributing a figure-like quality to land and a background-like quality to bodies of water, for instance, is thus but a mere cartographic convention. Indeed, on many nautical charts, the sea actually

> tends to assume the appearance which the land has on ordinary maps. [Although] the contour of the land is the same on a maritime chart as it is on a map of the usual type . . . when looking at such a map, say, of the Mediterranean, we may completely fail to see Italy. Instead we may see a strange figure, corresponding to the area of the Adriatic . . . which happens to have shape . . . [S]o long as the Mediterranean has shape, the area corresponding to Italy has no shape, and vice versa.[68]

Such maps may indeed look odd to us because they actually force us to effectively reverse the way we conventionally view land masses as figure-like and bodies of water as background-like.[69]

It might also be helpful at this point to reexamine our very notions of relevance and irrelevance. We usually regard relevance as simply a matter of logic, and consider what we inattend or disattend to intrinsically irrelevant or "extraneous."[70] And when reading experts write about teaching children how to identify "the main idea" of a text and distinguish "key" information from "minor details" in it,[71] one may get the impression that relevance is indeed objectively determinable.

Yet other than perhaps in analytic philosophy, legal scholarship, and mathematics, it is rarely logic alone that delineates the scope of our attention and concern and thereby determines what we consider relevant. It is never simply logic, for instance, that compels jurors to ignore inadmissible evidence presented in court. Indeed, as one can tell from comparing the front page of *The New York Times* with that of Rutgers University students' *The Daily Targum* on any given day, *nothing is inherently relevant* (or irrelevant, for that matter). Viewing anything as inherently relevant (and thereby "intrinsically foregroundable"),[72] therefore, effectively essentializes the merely conventional way in which we actually attribute relevance.

Given all this, one may also have to take the claim about the greater distractibility of "underfocused"[73] (or simply "unfocused")[74] people with so-called attention deficit disorder (ADD), which tacitly implies universal agreement on what constitutes a "distraction," with a grain of salt. After all, it may very well be that their attention span is simply shorter (that is, that they shift their attention somewhat more rapidly) than what we *conventionally* consider normal.[75] By the same token, it is quite possible that people with so-called obsessive-compulsive disorder (OCD), whom we regard as being "excessively" attentive to detail, as well as people with other forms of anxiety disorder whom we consider "hypervigilant,"[76] are but attentional deviants who simply disregard our effectively *conventional* attentional traditions and norms. And though the highly distinctive manner in which people with autism typically focus their attention[77] likewise makes them "appear to ignore relevant stimuli in favour of apparently meaningless" ones, we must keep in mind that "the decision which stimuli are relevant and which are meaningless"[78] is ultimately a *conventional* one.

Furthermore, as neurodiversity advocates point out, it is quite possible that people with autism simply have a distinctive attentional "style" rather than a "dysfunction"[79] and that we should therefore perhaps consider autism "a different type of information-processing system" rather than a disability.[80] In fact, claims Temple Grandin, noting that they are often exceptionally good at differentiating figure-like objects from their background-like surroundings (as exemplified by their ability to detect the constituent notes of musical chords or to spot familiar shapes "hidden" in perceptually complex configurations),[81] it is actually neurotypicals who are somehow "disabled" compared to them.[82] "I thought about [that] after 9/11," she adds,

> when news reports started coming out about how hard it is for . . . luggage inspectors to spot weapons . . . If [you sit] all day long staring at a video screen, pretty soon you'll have trouble separating out the form of a weapon from all the other junk that's packed in people's bags. The screen is too cluttered, and everything blurs together. But that might not be a problem for autistic people, and I think airports ought to try out some autistic people in that job.[83]

FOREGROUNDING

There is a famous scene in Alfred Hitchcock's *North by Northwest* in which Cary Grant stands on a deserted road waiting to meet the man for whom he believes he has been mistaken. As we watch him waiting, we catch a glimpse of a small, innocuous-looking crop duster flying in the background. A few minutes later we get a second glimpse of the plane, still in the background, and a man waiting for the bus on the other side of the road mutters: "Some of them crop-duster pilots get rich if they live long enough," then adding: "That's funny. That plane is dusting crops where there ain't no crops." It takes another long minute, however, before the plane reappears yet again, this time clearly in the foreground, and our hero realizes that it is actually heading in his direction, trying to kill him.

Like René Magritte's *The Muscles of the Sky* (see Plate 4), let alone the real-life September 11, 2001 attacks on the World Trade Center, the scene clearly challenges our conventional view of the sky as "essentially" background-like (see Figure 2.3). As it so compellingly demonstrates, since nothing is inherently background-like, anything can actually be perceptually pulled out of the background and become a figure-like object of attention in its own right. In order to be noticed, however, it first needs to be "foregrounded."

Foregrounding[84] calls for a fundamental "gestalt switch,"[85] which effectively involves reversing the conventional relation between figures and backgrounds by shifting our attention from the former to the latter.[86] In other words, it is essentially a process of *refocusing*, in which "*background becomes foreground, and foreground becomes background*,"[87] such as when we come to see life as "what happens to [us] while [we]'re busy making other plans"[88] and a chicken as a means for one egg to produce another.

Such *figure-ground reversal* ultimately involves turning the proverbial spotlight on what is usually ignored,[89] such as watching others' conventionally backgrounded feet instead of their typically foregrounded (and therefore highly self-monitored) face in order to access their unacknowledged feelings,[90] incorporating the conventionally backgrounded ringing of a cell phone into one's viola recital by playfully improvising on the ringtone,[91] or producing a documentary film spotlighting backup singers. (As Mark Olsen so succinctly put it in his aptly-titled film review "'20 Feet from Stardom' Moves Spotlight to the Background," "You've seen them but not noticed them. You've heard them but not listened to them.")[92] This is also true of any effort to turn a conventionally backgrounded (and therefore publicly invisible) issue into a focally attended "social problem."

Effectively defying our habitual, almost automatic[93] way of attending to the world which ultimately "strangles [our] awareness [by] limit[ing] us to [notic]ing only a fraction of what there is to be [noticed],"[94] foregrounding helps us make things "jum[p] out of the background enough to be perceived consciously rather than just being part of [our] surroundings."[95] And as so vividly exemplified by Alexandra Horowitz's writings about ordinary urban walks, in which she actually examines "what it is that I miss, every day, right in front of me, while walking around the block,"[96] that often involves *defamiliarization*,[97] a process of "making the familiar strange" by focusing our attention on the unmarked elements of our phenomenal world that we habitually take for granted and therefore inattend.[98] That is certainly true, for example, of street photography, of Horace Miner's satirical portrayal of ordinary American health and cleanliness habits as exotic "Nacirema" rituals,[99] as well as of Goffman's and Harold Garfinkel's efforts to unveil the taken-for-granted interactional order underlying everyday social life.[100] As someone once described such efforts, "Like the hidden face in the picture, it's hard not to see, once it's pointed out; Goffman not only sees it, but makes the rest of us see it too."[101]

Foregrounding helps *make the implicit explicit*,[102] as so compellingly exemplified by the story of Adam and Eve. As soon as they eat the fruits of the proverbial tree of knowledge or awareness, "the eyes of them both [a]re opened" (indeed, we often regard situations in which we suddenly become explicitly aware of something as "eye-openers"), and they immediately realize that they are naked,[103] something of which they were only tacitly cognizant before.

In order to help make the implicit explicit, foregrounding sometimes also involves simply naming the hitherto backgrounded, thus making it easier to attend to.[104] In fact, as implied in the following exchange, in Arthur Conan Doyle's story "Silver Blaze," between police inspector Gregory and Sherlock Holmes about the dog that did not bark on the night the horse disappeared, such an act can actually help foreground even non-events:

> "Is there any point to which you would wish to draw my attention?"
> "To the curious incident of the dog in the night-time."
> "The dog did nothing in the night-time."
> "That was the curious incident," remarked Sherlock Holmes.[105]

Naming non-occurrences is analogous, in a way, to the mental exercise of focusing visually on what Edmund Husserl so insightfully identified as "the between":[106]

> Look at a scene in nature imagining that the . . . spaces are solid. Instead of focusing on the substance, *look at the 'emptiness.'* If you are looking at a woodsy scene, ignore the trees and *concentrate on seeing the spaces between* the branches and leaves. . . . Let empty spaces . . . take on more importance.[107]

> [R]everse . . . *figure-and-ground awareness* so that you allow the figure to recede and . . . the background to move forward in your awareness. . . . *What was figure is now background* [and] *what was background is now figure* . . . [Allow] the side walls, the ceiling, and the floor to serve as figure and . . . the front wall to be seen as the background . . . highlighting peripheral awareness as figure and central awareness as ground.[108]

> [I]magine the space between your middle and ring fingers.[109]

Visual artists, in fact, are often given similar advice. Beginning photographers, for example, are thus instructed to

> *[e]mphasize the negative space* when taking a picture. *Learn to see the interval between visual elements* as figure. Position your camera in such a way as to make the interval between [objects] the integral part of your picture.[110]

By the same token, in her drawing classes, art instructor Betty Edwards trains her students to notice the shapes of the spaces formed between pieces of furniture in a room,[111] noting that they are "not just empty air" and just as important as the objects they happen to surround.[112] Challenging the "negative" character conventionally assigned to such spaces, a fellow art instructor likewise insists that they ought to be explicitly delineated.[113]

In fact, notes Edwards, while delineating figures, contours at the same time also delineate their surrounding backgrounds, thus providing them with a recognizable shape.[114] As she trains her students to notice the shapes of the spaces formed between pieces of furniture, she thus reminds them that a contour, after all, "is always the border of two things simultaneously,"[115] and that "[i]f you draw the edges of the spaces," you therefore "also will have drawn the chair, because it shares edges with the spaces."[116]

That idea, in fact, underlies the artistic production, dating back to antiquity,[117] of perceptually ambiguous configurations in which figures and backgrounds are effectively interchangeable.[118] Roger Shepard's *Beckoning Balusters* (see Figure 5.2),[119] in which the background-like gaps between the balusters also double as human-shaped figures in their own right, and Shigeo Fukuda's *Legs of Two Different Genders* (see Figure 5.3) in which the background-like spaces between the upward-pointing male legs also double as downward-pointing female legs, are classic examples of such "optical puns."[120]

So, indeed, are Sandro Del-Prete's *Secret between Fall Leaves* (see Plate 5), in which the seemingly empty space between the leaves also doubles as a naked body, and Rob Gonsalves's aptly-titled *A Change of Scenery* (see Plate 6), in which a figure-like silhouette of an urban skyline turns out to also double as the background-like space formed between the shreds of a torn drape. Even more spectacular in this regard are Maurits Escher's *Circle Limit IV*[121] and *Plane Filling II* (see Figure 5.4 for the latter), in which *virtually all the featured figure-like forms also double as one another's backgrounds*.[122] In assigning equal perceptual salience to each of those forms,[123] such works effectively challenge the figure-like or background-like "nature" of the various constituents of our phenomenal world.

CONCLUSION | 87

FIGURE 5.2 Balusters or women? Courtesy of Roger N. Shepard

Indeed, as so strikingly exemplified by the rise of the traditionally backgrounded double bass as a featured solo instrument in modern jazz[124] (or by Tom Stoppard's *Rosencrantz and Guildenstern Are Dead*, an entire play centered around minor, background-like characters from *Hamlet*, for that matter),[125] the impulse to foreground the conventionally backgrounded is a major feature of modernism.[126] The space between walls, floors, and ceilings, for instance, traditionally viewed as a "negative" element in design, is thus considered by modern architects a "positive" one,[127] and fewer urban designers today are likely to view spaces between buildings as mere void.[128]

Yet modernism calls attention not only to "empty" space but also to "empty" time, as so vividly manifested in modernist

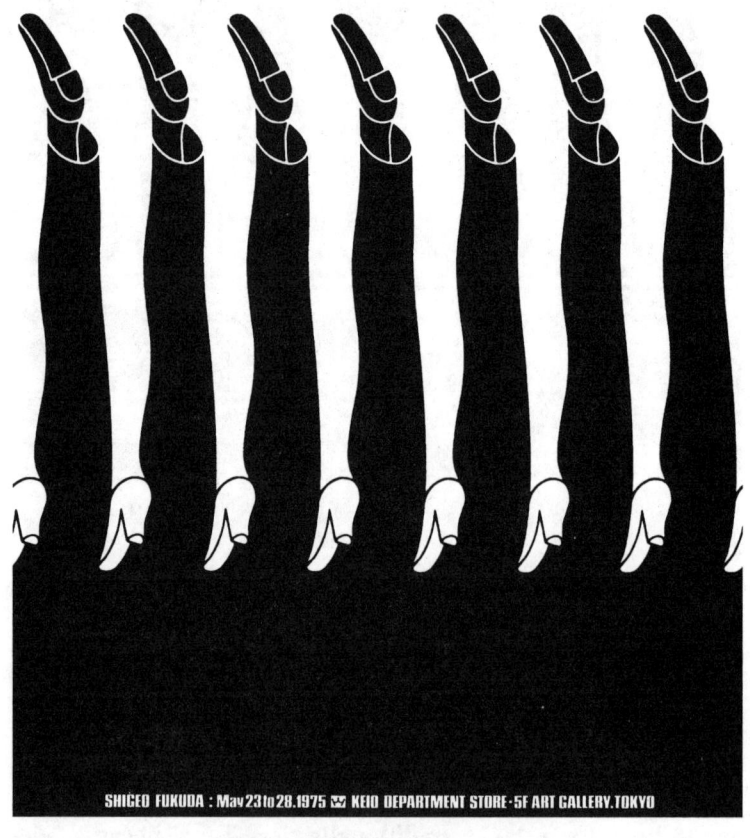

FIGURE 5.3 Men's or women's legs? © Shigeo Fukuda

CONCLUSION | 89

FIGURE 5.4 Figures or backgrounds? M. C. Escher's *Plane Filling II*. © 2014 The M. C. Escher Company—The Netherlands. All rights reserved

efforts to foreground silence. Stéphane Mallarmé's poems, in which conventionally backgrounded pauses are typographically foregrounded in the form of wider-than-average gaps between as well as within lines,[129] are classic examples of such efforts. Such heightened awareness of silence is also one of the most distinctive features of Jackson Mac Low's poems,[130] Samuel Beckett's and Harold Pinter's plays, as well as the music of Anton Webern and John Cage.

BEYOND FIGURE AND BACKGROUND

Even more epistemically disputable than our tendency to essentialize the seemingly intrinsic figure-like or background-like character of what we conventionally consider "figure" and "background," however, is the way we essentialize the very distinction between those two categories. According to the Gestalt theory of perception, our attention "is by nature directed toward objects" rather than toward the "intermediate spaces" between them,[131] and we therefore seem to "see things and not the holes between them."[132] Yet are those "objects" indeed inherently thing-like, and the "intermediate spaces" between them inherently hole-like? Could it be that we actually come to regard them as such only because we have been socialized, and therefore become habituated, to view them this way?

In the first place, what we consider "things" are not inherently more perceptually dominant than what we consider "nonthings."[133] Like Nino Rota's, Bernard Herrmann's, Philip Glass's, and Mozart's supposedly background-like soundtracks of *La Dolce Vita*, *Psycho*, *Koyaanisqatsi*, and *Elvira Madigan*, respectively, the spaces between the stems, flowers, and leaves in ikebana arrangements (see, for example, Plate 7)[134] contribute as much to the overall aesthetic effect as those supposedly thing-like elements themselves. Nor, for that matter, can one ignore "empty" spaces when designing a bracelet, a pair of scissors, or a ring.[135]

Furthermore, as so vividly demonstrated in Coles Phillips's "Fadeaway Girl" illustrations (see, for example, Plate 8 and Plate 9), although it is their envisioned contours that indeed help us mentally detach figures from their surrounding backgrounds, thus providing them with a thing-like quality, in the real world there is nothing actually separating figure-like objects from their background-like surroundings.[136] "I don't want to destroy any beloved delusions," points out art instructor Carl Purcell, "but the coloring books we grew up with lied!"[137] The contours delineating the "objects" we attend to and separating them from their "surroundings," which we habitually ignore, are but "an artistic convention to which we have become so accustomed that we overlook it,"[138] a mental device that we ourselves have created in order to underscore the supposedly discrete nature of what we envision as thing-like entities. In other words, they are mere products of our imagination.

Indeed, the very notion that the figure-like "things" that we notice are somehow detachable from their background-like "surroundings" that can therefore be ignored is by no means self-evident[139]—a central theme underlying the deliberate attempts by Frank Lloyd Wright and other "organic" architects to visually integrate buildings into their natural surroundings, as well as the modernist effort to allow art and its traditionally out-of-frame and therefore backgrounded non-artistic surroundings to literally interpenetrate each other. Thus, for example, in his "A Woman Seated beside a Vase of Flowers" (see Plate 10), not only does Edgar Degas defy the tradition of directing the viewer's attention to the very center of the painting,[140] as far away as possible from its non-artistic surroundings, he actually directs it (through the woman's gaze) to those conventionally backgrounded surroundings.[141] Such a glaring defiance of one of our most fundamental attentional conventions is even more dramatically exemplified in Piet Mondrian's "Lozenge Composition with Red, Gray, Blue, Yellow, and Black" (see Plate 11),[142] which almost begs the viewer to mentally "complete" the truncated geometrical forms beyond the canvas, let alone in Bill Mack's "Melody" (see Plate 12) and Pere Borrell del Caso's "Escaping Criticism" (see Plate 13), in which a sculpted or painted object is portrayed as literally flowing or stepping out of the "art" frame—a visually stunning effect also produced cinematically when a fictional character steps out of the screen into a movie theater in Woody Allen's film *The Purple Rose of Cairo*. It likewise manifests itself in Edward E. Cummings' poems that begin in the middle of a word within parentheses or end in the middle of a sentence,[143] as well as in the blurring of the distinction between performers and spectators in the theater (as so vividly exemplified in Luigi Pirandello's *Tonight We Improvise*, Clifford Odets's *Waiting for Lefty*, and Joseph Heller's *We Bombed in New Haven*).[144]

The ultimate expression of the modernist effort to integrate art and its conventionally backgrounded non-artistic surroundings, of course, are so-called "environments" that, in marked contrast to traditional paintings, for instance, are virtually unframed.[145] As such, there is indeed absolutely nothing around them that viewers can actually ignore as mere "background."

The truth, of course, is that we do not really have to choose between selective attention and open awareness, since they actually complement each other. Given the fact that our eyes, for example, contain cones, which enable us to focus our attention on figure-like objects, as well as rods, which allow us unfocused[146] access to their background-like surroundings, we are indeed hard-wired to utilize both modes of awareness. (And while the figure-and-ground model may indeed underlie the phenomenology of our conscious perception, it is the unfocused, "blank" stare[147] that actually captures the essence of unconscious perception.)[148] Using both our cone-dominant central vision and rod-dominant peripheral vision, we can thus attend to the world around us in *a focused as well as an unfocused* manner.

By alternating between those two effectively complementary modes of awareness we thus have more than just one way in which we can access our phenomenal world.[149] While eating curry, for example, we can thus discriminate the distinctive flavor of the coriander from that of the garlic, ginger, cumin, or turmeric, yet at the same time also savor the rich flavor of the sauce as a whole (just as we savor the sound of chords, as individual notes harmoniously blend with one another). In fact, the pleasure of eating that dish is probably enhanced by our ability to use both modes of accessing it.

Consider also, finally, the range of our (perceptual as well as conceptual) attention. Our mental "horizons,"[150] after all, are but figments of our minds, and we therefore need to be careful not to essentialize them. Indeed, as soon as we literally or figuratively turn our necks, their very position is immediately reconfigured.

Unlike people who actually lose their peripheral vision[151] (or horses with blinders), it is in fact we ourselves who essentially construct the narrow figurative tunnels to which we habitually confine our mental vision. Rather than resign ourselves to such a predicament, we can therefore "widen" our mental vision,[152] thus allowing ourselves access to what is ordinarily "hidden" from us, albeit in plain sight. After all, as James so wisely observed, by the way we attend to things we actually choose what sort of universe we ultimately inhabit.[153]

NOTES

Preface

1. Later published as *Patterns of Time in Hospital Life: A Sociological Perspective* (Chicago: University of Chicago Press, 1979), 126–30.
2. Eviatar Zerubavel, *Hidden Rhythms: Schedules and Calendars in Social Life* (Berkeley: University of California Press, 1985 [1981]), 19–30.
3. Eviatar Zerubavel, "Horizons: On the Sociomental Foundations of Relevance," *Social Research* 60 (1993): 397–413.
4. Eviatar Zerubavel, *Social Mindscapes: An Invitation to Cognitive Sociology* (Cambridge, MA: Harvard University Press, 1997), 35–52.
5. Eviatar Zerubavel, *The Elephant in the Room: Silence and Denial in Everyday Life* (New York: Oxford University Press, 2006), 65–68.

Chapter 1

1. Tom Brown, "Fill Your Senses, Light Up Your Life," *Reader's Digest* (August 1984): 153–54.
2. Joseph Jastrow, *The Subconscious* (Boston: Houghton, Mifflin & Co., 1906), 51.
3. See also Wolfgang Metzger, *Laws of Seeing* (Cambridge, MA: MIT Press, 2006 [1936]), 1; Max H. Bazerman and Dolly Chugh,

"Bounded Awareness: Focusing Failures in Negotiation," in Leigh L. Thompson (ed.), *Negotiation Theory and Research* (New York: Psychology Press, 2006), 11; Alexandra Horowitz, *On Looking: Eleven Walks with Expert Eyes* (New York: Scribner, 2013), 2.
4. Christopher Chabris and Daniel Simons, *The Invisible Gorilla and Other Ways Our Intuition Deceives Us* (London: HarperCollins, 2010), 16.
5. P. Sven Arvidson, *The Sphere of Attention: Context and Margin* (Dordrecht, the Netherlands: Springer, 2006), 13.
6. Arien Mack and Irvin Rock, *Inattentional Blindness* (Cambridge, MA: MIT Press, 1998), 14, 227–28; Arien Mack, "Inattentional Blindness: Looking Without Seeing," *Current Directions in Psychological Science* 12 (2003): 180.
7. William James, *The Principles of Psychology* (Cambridge, MA: Harvard University Press, 1983 [1890]), 380.
8. John Dewey, *Psychology*, 3rd ed. (New York: Harper and Brothers, 1893), 134.
9. Herbert J. Schlesinger, "Cognitive Attitudes in Relation to Susceptibility to Interference," *Journal of Personality* 22 (1954): 356.
10. Jonathan Crary, *Suspensions of Perception: Attention, Spectacle, and Modern Culture* (Cambridge, MA: MIT Press, 1999), 1.
11. Ibid., 24, 39.
12. Aldous Huxley, "The Doors of Perception," in *The Doors of Perception and Heaven and Hell* (New York: HarperCollins, 2009 [1954]), 23. See also 24, 144–45; Henri Bergson, "'Phantasms of the Living' and 'Psychical Research,'" in *Mind-Energy: Lectures and Essays* (New York: Henry Holt, 1920 [1913]), 95–96.
13. Constance Grauds and Doug Childers, *The Energy Prescription: Give Yourself Abundant Vitality with the Wisdom of America's Leading Natural Pharmacist* (New York: Bantam Books, 2005), 185–86.
14. Robert Ornstein, *Meditation and Modern Psychology* (Los Altos, CA: Malor Books, 2008 [1971]), 8–9, 14, 16, 26–27, 29, 35; Livia Kohn, "Meditation and Visualization," in Fabrizio Pregadio (ed.), *The Encyclopedia of Taoism*, vol. 1 (New York: Routledge, 2008), 118; James H. Austin, *Meditating Selflessly: Practical Neural Zen* (Cambridge, MA: MIT Press, 2011), 8. See also Thomas Keefe, "Meditation and Social Work Treatment," in Francis J. Turner (ed.), *Social Work Treatment: Interlocking Theoretical Approaches*, 4th ed. (New York: Free Press, 1996),

435; Bruce R. Dunn et al., "Concentration and Mindfulness Meditations: Unique Forms of Consciousness?" *Applied Psychophysiology and Biofeedback* 24 (1999): 147–65; James H. Austin, *Zen-Brain Reflections: Reviewing Recent Developments in Meditation and States of Consciousness* (Cambridge, MA: MIT Press, 2006), 30; Jeffrey B. Rubin, "Deepening Listening: The Marriage of Buddha and Freud," in Uwe P. Gielen et al. (eds.), *Principles of Multicultural Counseling and Therapy* (New York: Routledge, 2008), 378; John J. Pilch, *Flights of the Soul: Visions, Heavenly Journeys, and Peak Experiences in the Biblical World* (Grand Rapids, MI: William B. Eerdmans, 2011), 178.
15. Paul L. Wachtel, "Conceptions of Broad and Narrow Attention," *Psychological Bulletin* 68 (1967): 417.
16. See, for example, Ulric Neisser, *Cognitive Psychology* (New York: Appleton-Century-Crofts, 1967), 96; Michael Posner, "Psychobiology of Attention," in Michael S. Gazzaniga and Colin Blakemore (eds.), *Handbook of Psychobiology* (New York: Academic Press, 1975), 443; Harold E. Pashler, *The Psychology of Attention* (Cambridge, MA: MIT Press, 1999), 2; Crary, *Suspensions of Perception*, 24; Chabris and Simons, *The Invisible Gorilla*, 9, 13.
17. Crary, *Suspensions of Perception*, 24. See also Rubin, "Deepening Listening," 378.
18. Dewey, *Psychology*, 134.
19. See also Ernest G. Schachtel, *Metamorphosis: On the Development of Affect, Perception, Attention, and Memory* (New York: Basic Books, 1959), 253–54.
20. Wilhelm Dilthey, *Selected Works, vol. I: Introduction to the Human Sciences* (Princeton, NJ: Princeton University Press, 1989 [1883]), 314.
21. See, for example, Jastrow, *The Subconscious*, 50; Richard Jung, "Correlation of Bioelectrical and Autonomic Phenomena with Alterations of Consciousness and Arousal in Man," in J. F. Delafresnaye (ed.), *Brain Mechanisms and Consciousness* (Springfield, IL: Charles C. Thomas, 1954), 313; Michael I. Posner et al., "Attention and the Detection of Signals," *Journal of Experimental Psychology: General* 109 (1980): 172; David LaBerge, *Attentional Processing: The Brain's Art of Mindfulness* (Cambridge, MA: Harvard University Press, 1995), 27; Jeanne Townsend and Eric Courchesne, "Parietal Damage and Narrow 'Spotlight' Spatial Attention," *Journal of Cognitive Neuroscience*

6 (1994): 222; Kyle R. Cave and Narcisse P. Bichot, "Visuospatial Attention: Beyond a Spotlight Model," *Psychonomic Bulletin and Review* 6 (1999): 204; Diego Fernandez-Duque and Mark Johnson, "Attention Metaphors: How Metaphors Guide the Cognitive Psychology of Attention," *Cognitive Science* 23 (1999): 93–102.
22. Raúl Hernández-Peón, "Psychiatric Implications of Neurophysiological Research," *Bulletin of the Menninger Clinic* 28 (1964): 167.
23. Steven S. Hsiao and Francisco Vega-Bermudez, "Attention in the Somatosensory System," in Randall J. Nelson (ed.), *The Somatosensory System: Deciphering the Brain's Own Body Image* (Boca Raton, FL: CRC Press LLC, 2001), 213. See also 202.
24. Emile Durkheim, "Individual and Collective Representations," in *Sociology and Philosophy* (New York: Free Press, 1974 [1898]), 21.
25. Alfred Binet and Charles Féré, *Animal Magnetism*, 5th ed. (London: Kegan Paul, Trench, Trübner, and Co., 1905 [1887]), 319.
26. Chabris and Simons, *The Invisible Gorilla*, 38.
27. See also Drew Leder, *The Absent Body* (Chicago: University of Chicago Press, 1990), 38.
28. Daniel J. Simons and Christopher F. Chabris, "Gorillas in Our Midst: Sustained Inattentional Blindness for Dynamic Events," *Perception* 28 (1999): 1059–74; Chabris and Simons, *The Invisible Gorilla*, 5–7, http://www.theinvisiblegorilla.com/videos.html (accessed on September 9, 2012). See also Ulric Neisser and Robert Becklen, "Selective Looking: Attending to Visually Specified Events," *Cognitive Psychology* 7 (1975): 480–94; Robert Becklen and Daniel Cervone, "Selective Looking and the Noticing of Unexpected Events," *Memory and Cognition* 11 (1983): 601–08; Daniel Memmert, "The Effects of Eye Movements, Age, and Expertise on Inattentional Blindness," *Consciousness and Cognition* 15 (2006): 620–27.
29. James, *The Principles of Psychology*, 430.
30. Mack and Rock, *Inattentional Blindness*.
31. See, for example, Ronald A. Rensink et al., "To See or Not To See: The Need for Attention to Perceive Changes in Scenes," *Psychological Science* 8 (1997): 373; Mack, "Inattentional Blindness," 180; Daniel Memmert and Philip Furley, "'I Spy with My Little Eye!': Breadth of Attention, Inattentional Blindness, and Tactical Decision Making in Team Sports," *Journal of Sport and*

Exercise Psychology 29 (2007): 365; Paola Bressan and Silvia Pizzighello, "The Attentional Cost of Inattentional Blindness," *Cognition* 106 (2008): 371.
32. Jastrow, *The Subconscious*, 52.
33. Tom Griffiths and Cathleen Moore, "A Matter of Perception," *Aquatics International* (November/December 2004), http://www.aquaticsintl.com/2004/nov/0411_rm.html (accessed on August 30, 2012).
34. See also Chabris and Simons, *The Invisible Gorilla*, 6–7, 38–39.
35. See, for example, Mack and Rock, *Inattentional Blindness*, 215–21.
36. Hsiao and Vega-Bermudez, "Attention in the Somatosensory System," 197. See also 198, 202; Leder, *The Absent Body*; Mack and Rock, *Inattentional Blindness*, 223–25.
37. Peggy DesAutels, "Gestalt Shifts in Moral Perception," in Larry May et al. (eds.), *Mind and Morals: Essays on Cognitive Science and Ethics* (Cambridge, MA: MIT Press, 1996), 133.
38. See, for example, Ronald S. Friedman et al., "Attentional Priming Effects on Creativity," *Creativity Research Journal* 15 (2003): 277–86. See also Hernández-Peón, "Psychiatric Implications of Neurophysiological Research," 167; Richard E. Nisbett, *The Geography of Thought: How Asians and Westerners Think Differently . . . and Why* (New York: Free Press, 2003).
39. Wayne C. Burkan, *Wide-Angle Vision: Beat Your Competition by Focusing on Fringe Competitors, Lost Customers, and Rogue Employees* (New York: John Wiley & Sons, 1996); George S. Day and Paul J. H. Schoemaker, *Peripheral Vision: Detecting the Weak Signals That Will Make or Break Your Company* (Boston: Harvard Business School Press, 2006); Karen A. Cerulo, *Never Saw It Coming: Cultural Challenges to Envisioning the Worst* (Chicago: University of Chicago Press, 2006).
40. See also Thomas H. Davenport and John C. Beck, *The Attention Economy: Understanding the New Currency of Business* (Boston: Harvard Business School Press, 2001), 58.
41. See, for example, Donald E. Broadbent, *Perception and Communication* (Oxford: Pergamon Press, 1958).
42. Asia Friedman, *Blind to Sameness: Sexpectations and the Social Construction of Male and Female Bodies* (Chicago: University of Chicago Press, 2013), 29. See also Hernández-Peón, "Psychiatric Implications of Neurophysiological Research," 167;

Fernandez-Duque and Johnson, "Attention Metaphors," 88–92; Crary, *Suspensions of Perception*, 41.
43. Friedman, *Blind to Sameness*, 29.
44. See, for example, Charles W. Mills, *The Racial Contract* (Ithaca, NY: Cornell University Press, 1997), 18, 93, 97. See also Robert N. Proctor, "Agnotology: A Missing Term to Describe the Cultural Production of Ignorance (and Its Study)," in Robert N. Proctor and Londa Schiebinger (eds.), *Agnotology: The Making and Unmaking of Ignorance* (Stanford, CA: Stanford University Press, 2008), 1–33.
45. Anna Freud, *The Ego and the Mechanisms of Defence* (London: Karnak Books, 1993 [1936]), 89. See also 69–92, 174.
46. Elizabeth F. Howell, *The Dissociative Mind* (New York: Routledge, 2005), 97.
47. Harry S. Sullivan, *Clinical Studies in Psychiatry* (New York: W. W. Norton, 1956). See also Steven N. Gold and Gonzalo Bacigalupe, "Interpersonal and Systemic Theories of Personality," in David F. Barone et al. (eds.), *Advanced Personality* (New York: Springer, 1998), 63.
48. Jennifer J. Freyd and Pamela Birrell, *Blind to Betrayal: Why We Fool Ourselves We Aren't Being Fooled* (Hoboken, NJ: John Wiley & Sons, 2013), x. See also Jennifer J. Freyd, *Betrayal Trauma: The Logic of Forgetting Childhood Abuse* (Cambridge, MA: Harvard University Press, 1996).
49. Noga Zerubavel, "Restricted Awareness in Intimate Partner Violence: The Effect of Childhood Sexual Abuse and Fear of Abandonment" (Ph.D. diss., Miami University, 2013).
50. Gresham M. Sykes and David Matza, "Techniques of Neutralization: A Theory of Delinquency," *American Sociological Review* 22 (1957): 668.
51. Chabris and Simons, *The Invisible Gorilla*, 38.
52. Eviatar Zerubavel, *Social Mindscapes: An Invitation to Cognitive Sociology* (Cambridge, MA: Harvard University Press, 1997), 36.
53. See also Eviatar Zerubavel, *The Fine Line: Making Distinctions in Everyday Life* (Chicago: University of Chicago Press, 1993 [1991]).
54. Zerubavel, *Social Mindscapes*, 35–52.
55. Eviatar Zerubavel, "Horizons: On the Sociomental Foundations of Relevance," *Social Research* 60 (1993): 398; Zerubavel, *Social Mindscapes*.

Chapter 2

1. Magdalen D. Vernon, *The Psychology of Perception* (Harmondsworth, UK: Penguin, 1962), 42. See also 40.
2. See also Magdalen D. Vernon, *Visual Perception* (Ann Arbor: University of Michigan Press, 1937), 91; James J. Gibson, "Perception of Distance and Space in the Open Air," in David C. Beardslee and Michael Wertheimer (eds.), *Readings in Perception* (Princeton, NJ: D. Van Nostrand Company, 1958 [1946]), 420; Alan Watts, *The Two Hands of God: The Myths of Polarity* (Collier Books, 1969), 21; Richard D. Zakia, *Perception and Imaging*, 2nd ed. (Boston: Focal Press, 2002), 4.
3. William James, *The Principles of Psychology* (Cambridge, MA: Harvard University Press, 1983 [1890]), 381; Joseph Jastrow, *The Subconscious* (Boston: Houghton, Mifflin & Co., 1906), 52; Edward B. Titchener, *A Text-Book of Psychology* (New York: Macmillan, 1910), 266.
4. Edgar Rubin, *Visuell Wahrgenommene Figuren: Studien in Psychologischer Analyse* (Copenhagen: Gyldendalske Boghandel, 1921 [1915]), 1–101. See also Jörgen L. Pind, "Figure and Ground at 100," *The Psychologist* 25, no. 1 (January 2012): 90–91, http://www.psy.ku.dk/om/Historie/figure_and_ground_at_100/JLPind-Psychologist.pdf/ (accessed on October 1, 2012).
5. Edgar Rubin, "Figure and Ground," in David C. Beardslee and Michael Wertheimer (eds.), *Readings in Perception* (Princeton, NJ: D. Van Nostrand Company, 1958 [1915]), 199. See also Harry Helson, "The Psychology of Gestalt," *American Journal of Psychology* 36 (1925): 495–96.
6. See also Maurice Merleau-Ponty, *Phenomenology of Perception* (New York: Humanities Press, 1962 [1945]), 13, 24; Daniel Kahneman, *Attention and Effort* (Englewood Cliffs, NJ: Prentice-Hall, 1973), 76.
7. See also Wolfgang Metzger, *Laws of Seeing* (Cambridge, MA: MIT Press, 2006 [1936]), 3; J. Brian Subirana-Vilanova and Whitman Richards, "Attentional Frames, Frame Curves and Figural Boundaries: The Inside/Outside Dilemma," *Vision Research* 36 (1996): 1499.
8. See also Rudolf Arnheim, *Visual Thinking* (Berkeley: University of California Press, 1969), 284.

9. See, for example, Stephen Kern, *The Culture of Time and Space 1880–1918* (Cambridge, MA: Harvard University Press, 1983), 160.
10. See also Gyorgy Kepes, *Language of Vision* (Chicago: Paul Theobald, 1951), 69.
11. See, for example, John Green, *Ballet Class Coloring Book* (Mineola, NY: Dover, 1998); John Green, *The Language of Flowers Coloring Book* (Mineola, NY: Dover, 2003).
12. James J. Gibson, *The Perception of the Visual World* (Westport, CT: Greenwood Press, 1950), 38. See also Metzger, *Laws of Seeing*, 3–4, 6, 13; Merleau-Ponty, *Phenomenology of Perception*, 15.
13. See, for example, Helson, "The Psychology of Gestalt," 496. See also Ernst H. Gombrich, *Art and Illusion: A Study in the Psychology of Pictorial Representation* (London: Phaidon, 2002 [1960]), 191; Trudy Friend, *Landscape: Problems and Solutions* (Newton Abbot, UK: David & Charles, 2004), 16.
14. See, for example, Rubin, "Figure and Ground," 196; Kurt Koffka, *Principles of Gestalt Psychology* (New York: Harcourt, Brace & World, 1935), 44, 208–09; Wolfgang Köhler, *Gestalt Psychology: An Introduction to New Concepts in Modern Psychology* (New York: Mentor Books, 1947), 121; Gibson, *The Perception of the Visual World*, 38; Roberto Casati and Achille C. Varzi, *Holes and Other Superficialities* (Cambridge, MA: MIT Press, 1994), 168. See also Gaetano Kanizsa, "Perception, Past Experience, and the 'Impossible Experiment,'" in *Organization in Vision: Essays on Gestalt Perception* (New York: Praeger, 1979 [1969]), 47; Friend, *Landscape*, 28, 94, 113, 118.
15. See also Roger Trancik, *Finding Lost Space: Theories of Urban Design* (New York: John Wiley & Sons, 1986), 98–99.
16. See, for example, Koffka, *Principles of Gestalt Psychology*, 186. See also Rubin, "Figure and Ground," 197.
17. See also Metzger, *Laws of Seeing*, 7.
18. See also Titchener, *A Text-Book of Psychology*, 266–67, 276–79; Ernest G. Wever, "Attention and Clearness in the Perception of Figure and Ground," *American Journal of Psychology* 40 (1928): 74; Vernon, *The Psychology of Perception*, 42; Charles Goodwin and Alessandro Duranti, "Rethinking Context: An Introduction," in Alessandro Duranti and Charles Goodwin (eds.), *Rethinking Context: Language as an Interactive Phenomenon* (Cambridge: Cambridge University Press, 1992), 10.

19. See also Titchener, *A Text-Book of Psychology*, 278–79; Rubin, "Figure and Ground," 197; Kurt Koffka, "Perception: An Introduction to the *Gestalt-Theorie*," *Psychological Bulletin* 19 (1922): 557, 566; Koffka, *Principles of Gestalt Psychology*, 186, 194.
20. See also Rudolf Arnheim, *Art and Visual Perception: A Psychology of the Creative Eye* (Berkeley: University of California Press, 1967 [1954]), 63; Arnheim, *Visual Thinking*, 286.
21. See also Leonard Shlain, *The Alphabet versus the Goddess: The Conflict between Word and Image* (New York: Penguin Arkana, 1998), 24; Michael Kubovy and David Van Valkenburg, "Auditory and Visual Objects," *Cognition* 80 (2001): 102, 104.
22. See, for example, Eva Wong and Naomi Weisstein, "Sharp Targets Are Detected Better against a Figure, and Blurred Targets Are Detected Better against a Background," *Journal of Experimental Psychology: Human Perception and Performance* 9 (1983): 194; Glyn W. Humphreys et al., "Attending But Not Seeing: The 'Other Race' Effect in Face and Person Perception Studied through Change Blindness," *Visual Cognition* 12 (2005): 249–62; Veronica Mazza et al., "Foreground-Background Segmentation and Attention: A Change Blindness Study," *Psychological Research* 69 (2005): 201–10. See also Vernon, *The Psychology of Perception*, 42.
23. Anton Ehrenzweig, *The Hidden Order of Art: A Study in the Psychology of Artistic Imagination* (Berkeley: University of California Press, 1971 [1967]), 11; Koffka, *Principles of Gestalt Psychology*, 186–89. See also Helson, "The Psychology of Gestalt," 495–96.
24. Rubin, "Figure and Ground," 194, 201. See also Koffka, "Perception," 556–57, 565; Metzger, *Laws of Seeing*, 3, 6, 13; Köhler, *Gestalt Psychology*, 107, 110; Arnheim, *Art and Visual Perception*, 223, 230–31; Julian E. Hochberg, *Perception* (Englewood Cliffs, NJ: Prentice-Hall, 1964), 59; Arnheim, *Visual Thinking*, 284; Kanizsa, "Perception, Past Experience, and the 'Impossible Experiment,'" 47; Douglas R. Hofstadter, *Gödel, Escher, Bach: An Eternal Golden Braid* (New York: Basic Books, 1979), 68; Trancik, *Finding Lost Space*, 99; Mary A. Peterson and Bradley S. Gibson, "The Initial Identification of Figure-Ground Relationships: Contributions from Shape Recognition Processes," *Bulletin of the Psychonomic Society* 29 (1991): 199; Gordon C. Baylis

and Ellison M. Cale. "The Figure Has a Shape, But the Ground Does Not: Evidence from a Priming Paradigm," *Journal of Experimental Psychology: Human Perception and Performance* 27 (2001): 633–43; Nava Rubin, "Figure and Ground in the Brain," *Nature Neuroscience* 4 (2001): 857; Brita Wårvik, "What Is Foregrounded in Narratives? Hypotheses for the Cognitive Basis of Foregrounding," in Tuija Virtanen (ed.), *Approaches to Cognition Through Text and Discourse* (Berlin: Mouton de Gruyter, 2004), 105; Roy Sorensen, *Seeing Dark Things: The Philosophy of Shadows* (New York: Oxford University Press, 2008), 186; Pind, "Figure and Ground at 100," 91.

25. See also Koffka, "Perception," 556–57; Kanizsa, "Perception, Past Experience, and the 'Impossible Experiment,'" 47; Wårvik, "What Is Foregrounded in Narratives?" 105.

26. See also Rubin, "Figure and Ground," 194–95; Peterson and Gibson, "The Initial Identification of Figure-Ground Relationships," 199.

27. Rubin, "Figure and Ground," 194–95. See also Koffka, "Perception," 557; Elizabeth S. Spelke et al., "The Development of Object Perception," in Stephen M. Kosslyn and Daniel N. Osherson (eds.), *Visual Cognition: An Invitation to Cognitive Science*, 2nd ed., vol. 2 (Cambridge, MA: MIT Press, 1995), 305; Jon Driver and Gordon C. Baylis, "Edge-Assignment and Figure-Ground Segmentation in Short-Term Visual Matching," *Cognitive Psychology* 31 (1996): 252; Baylis and Cale, "The Figure Has a Shape, But the Ground Does Not"; Xiaofeng Ren et al., "Figure/Ground Assignment in Natural Images," *Lecture Notes in Computer Science* 3952 (2006): 614–15.

28. Aron Gurwitsch, *The Field of Consciousness* (Pittsburgh, PA: Duquesne University Press, 1964), 111. See also Koffka, *Principles of Gestalt Psychology*, 181–82; Heinz Werner and Seymour Wapner, "Toward a General Theory of Perception," in David C. Beardslee and Michael Wertheimer (eds.), *Readings in Perception* (Princeton, NJ: D. Van Nostrand Company, 1958 [1952]), 506–07; Arnheim, *Art and Visual Perception*, 217; Kahneman, *Attention and Effort*, 76–78.

29. Rudolf Arnheim, "The Perception of Maps," in *New Essays on the Psychology of Art* (Berkeley: University of California Press, 1986 [1976]), 199.

30. Anton Ehrenzweig, *The Psycho-Analysis of Artistic Vision and Hearing: An Introduction to a Theory of Unconscious Perception* (New York: The Julian Press, 1953), 26.
31. See also Casati and Varzi, *Holes and Other Superficialities*, 168.
32. See also Koffka, *Principles of Gestalt Psychology*, 209; Metzger, *Laws of Seeing*, 4; Köhler, *Gestalt Psychology*, 110.
33. Metzger, *Laws of Seeing*, 4.
34. Merleau-Ponty, *Phenomenology of Perception*, 13; Arnheim, *Art and Visual Perception*, 217, 223; Casati and Varzi, *Holes and Other Superficialities*, 160–61; Subirana-Vilanova and Richards, "Attentional Frames, Frame Curves and Figural Boundaries," 1493.
35. See, for example, Yael Zerubavel, "Space Metaphors in Modern Israeli Culture" (Paper presented at the Annual Meeting of the Association for Jewish Studies, Chicago, 2012).
36. Gurwitsch, *The Field of Consciousness*, 111.
37. Rubin, *Visuell Wahrgenommene Figuren*, Figure 3.
38. This image in fact dates back to the eighteenth century. See, for example, Zakia, *Perception and Imaging*, 1, http://gallica.bnf.fr/ark:/12148/btv1b8412334s/f1.highres (accessed on November 7, 2011). See also http://books.google.com/books?id=XXEFAAAAQAAJ&printsec=frontcover&dq=conchological+dictionary&hl=en&ei=vVuHToevJKL10gHe4M38Dw&sa=X&oi=book_result&ct=result&resnum=1&ved=0CDUQ6AEwAA#v=onepage&q&f=false (accessed on October 1, 2011).
39. Hochberg, *Perception*, 83.
40. See also Koffka, *Principles of Gestalt Psychology*, 201; Vernon, *The Psychology of Perception*, 42–43.
41. See also Zakia, *Perception and Imaging*, 12.
42. See, for example, David Katz, *The World of Touch* (Hillsdale, NJ: Lawrence Erlbaum Associates, 1989 [1925]), 61.
43. See, for example, Pierre L. Divenyi and Ira J. Hirsh, "Some Figural Properties of Auditory Patterns," *Journal of the Acoustical Society of America* 64 (1978): 1369–85; Albert S. Bregman, *Auditory Scene Analysis: The Perceptual Organization of Sound* (Cambridge, MA: MIT Press, 1990), 141. See also Donald E. Broadbent, *Perception and Communication* (Oxford: Pergamon Press, 1958), 60.

44. Bregman, *Auditory Scene Analysis*, 40. See also Paul Saenger, *Space Between Words: The Origins of Silent Reading* (Stanford, CA: Stanford University Press, 1997).
45. Ulric Neisser, *Cognitive Psychology* (New York: Appleton-Century-Crofts, 1967), 189; Jonathan Harrington and Steve Cassidy, *Techniques in Speech Acoustics* (Dordrecht, The Netherlands: Kluwer Academic Publishers, 1999), 126–28.
46. Eviatar Zerubavel, *The Fine Line: Making Distinctions in Everyday Life* (Chicago: University of Chicago Press, 1993 [1991]), 80.
47. Donald A. Hodges and David C. Sebald, *Music in the Human Experience: An Introduction to Music Psychology* (New York: Routledge, 2011), 133.
48. Albert S. Bregman and Jeffrey Campbell, "Primary Auditory Stream Segregation and Perception of Order in Rapid Sequences of Tones," *Journal of Experimental Psychology* 89 (1971): 244. See also Stephen McAdams and Albert Bregman, "Hearing Musical Streams," *Computer Music Journal* 3, no. 4 (December 1979): 26–43, 60; Bregman, *Auditory Scene Analysis*, 18.
49. McAdams and Bregman, "Hearing Musical Streams," 28. See also Bregman, *Auditory Scene Analysis*, 9–10, 18; Renaud Brochard et al., "Perceptual Organization of Complex Auditory Sequences: Effect of Number of Simultaneous Subsequences and Frequency Separation," *Journal of Experimental Psychology: Human Perception and Performance* 25 (1999): 1757.
50. Bregman, *Auditory Scene Analysis*, 138.
51. See also Koffka, *Principles of Gestalt Psychology*, 201; E. Colin Cherry, "Some Experiments on the Recognition of Speech, with One and with Two Ears," *Journal of the Acoustical Society of America* 25 (1953): 976; Kahneman, *Attention and Effort*, 79.
52. Philip Tagg, "Accompaniment," in John Shepherd et al. (eds.), *Continuum Encyclopedia of Popular Music of the World*, vol. II: *Performance and Production* (London: Continuum, 2003), 628. See also Hofstadter, *Gödel, Escher, Bach*, 70; Philip Tagg, "Melody and Accompaniment," http://www.andrelambert.org/uqam/principes/melodaccUS.pdf (accessed on August 1, 2012).
53. See also Ehrenzweig, *The Psycho-Analysis of Artistic Vision and Hearing*, 40.
54. See also Tagg, "Melody and Accompaniment," 18.
55. Tagg, "Accompaniment," 628.

56. See also Harris M. Berger, *Metal, Rock, and Jazz: Perception and the Phenomenology of Musical Experience* (Middletown, CT: Wesleyan University Press, 1999), 126.
57. Tagg, "Accompaniment," 628. See also David Horn and Stan Hawkins, "Backing," in John Shepherd et al. (eds.), *Continuum Encyclopedia of Popular Music of the World*, vol. II: *Performance and Production* (London: Continuum, 2003), 632.
58. Joseph Lanza, *Elevator Music: A Surreal History of Muzak, Easy-Listening, and Other Moodsong*, rev. ed. (Ann Arbor: University of Michigan Press, 2004), 3, 18, 73.
59. Ibid., 3. See also 18.
60. Joseph Lanza, "The Sound of Cottage Cheese (Why Background Music Is the Real World Beat!)" *Performing Arts Journal* 13, no. 3 (September 1991): 50, 43.
61. Lanza, *Elevator Music*, 19; Lanza, "The Sound of Cottage Cheese," 45.
62. Lanza, *Elevator Music*, 59.
63. See, for example, Ronald S. Friedman et al., "Attentional Priming Effects on Creativity," *Creativity Research Journal* 15 (2003): 277–86. See also Raúl Hernández-Peón, "Psychiatric Implications of Neurophysiological Research," *Bulletin of the Menninger Clinic* 28 (1964): 167.
64. Eviatar Zerubavel, *Social Mindscapes: An Invitation to Cognitive Sociology* (Cambridge, MA: Harvard University Press, 1997), 36.
65. See, for example, Goodwin and Duranti, "Rethinking Context: An Introduction," 9–10.
66. On "moral attention," see Zerubavel, *Social Mindscapes*, 39–40.
67. Carol Gilligan, "Moral Orientation and Moral Development," in Eva F. Kittay and Diana T. Meyers (eds.), *Women and Moral Theory* (Totowa, NJ: Rowman & Littlefield, 1987), 23. See also 22, 25; Peggy DesAutels, "Gestalt Shifts in Moral Perception," in Larry May et al. (eds.), *Mind and Morals: Essays on Cognitive Science and Ethics* (Cambridge, MA: MIT Press, 1996), 133.
68. See, for example, Eviatar Zerubavel, *Time Maps: Collective Memory and the Social Shape of the Past* (Chicago: University of Chicago Press, 2003), 25–34.
69. Ashley Parker and Michael Barbaro, "Romney Takes Analytic Approach to Campaign Chaos," *New York Times*, February 28, 2012, A16.

70. Paul J. Hopper and Sandra A. Thompson, "Transitivity in Grammar and Discourse," *Language* 56 (1980): 280.
71. Wårvik, "What Is Foregrounded in Narratives?" 99.
72. Erving Goffman, "Fun in Games," in *Encounters: Two Studies in the Sociology of Interaction* (Indianapolis, IN: Bobbs-Merrill, 1961), 25; Erving Goffman, *Frame Analysis: An Essay on the Organization of Experience* (New York: Harper & Row, 1974), 201–46.
73. Gregory Bateson, "A Theory of Play and Fantasy," in *Steps to an Ecology of Mind* (New York: Ballantine Books, 1972 [1955]), 177–93; Goffman, *Frame Analysis*.
74. James, *The Principles of Psychology*, 432.
75. Bateson, "A Theory of Play and Fantasy," 187. See also Arnheim, *Art and Visual Perception*, 231.
76. See also Goffman, *Frame Analysis*, 208; Bregman, *Auditory Scene Analysis*, 701.
77. See also Linda Waugh, "Marked and Unmarked: A Choice between Unequals in Semiotic Structure," *Semiotica* 38 (1982): 302. On the latter distinction, see, for example, Nikolai S. Trubetzkoy, *Principles of Phonology* (Berkeley: University of California Press, 1969 [1939]), 146–47; Waugh, "Marked and Unmarked"; Eviatar Zerubavel, *The Seven-Day Circle: The History and Meaning of the Week* (Chicago: University of Chicago Press, 1989 [1985]), 116–17; Wayne H. Brekhus, "Social Marking and the Mental Coloring of Identity: Sexual Identity Construction and Maintenance in the United States," *Sociological Forum* 11 (1996): 497–522; Ross Chambers, "The Unexamined," in Mike Hill (ed.), *Whiteness: A Critical Reader* (New York: New York University Press, 1997), 187–203; Wayne H. Brekhus, "A Sociology of the Unmarked: Redirecting Our Focus," *Sociological Theory* 16 (1998): 34–51; Eviatar Zerubavel, "The Social Marking of the Past: Toward a Socio-Semiotics of Memory," in Roger Friedland and John Mohr (eds.), *Matters of Culture: Cultural Sociology in Practice* (Cambridge: Cambridge University Press, 2004), 186–87.
78. See, for example, Henri Hubert, "A Brief Study of the Representation of Time in Religion and Magic," in *Essay on Time* (Oxford: Durkheim Press, 1999 [1905]), 43–91.
79. See also Allan V. Horwitz, "Normality," *Contexts* 7, no.1 (2008): 70–71.

80. See also Waugh, "Marked and Unmarked," 300.
81. Zerubavel, *The Seven-Day Circle*, 113–14. See also Eviatar Zerubavel, *Hidden Rhythms: Schedules and Calendars in Social Life* (Berkeley: University of California Press, 1985 [1981]), 115–16.
82. See Eviatar Zerubavel, *Ancestors and Relatives: Genealogy, Identity, and Community* (New York: Oxford University Press, 2011), 63.
83. Rubin, "Figure and Ground," 202; Koffka, *Principles of Gestalt Psychology*, 191; Hochberg, *Perception*, 86; Casati and Varzi, *Holes and Other Superficialities*, 159; Pind, "Figure and Ground at 100," 91.
84. See also Waugh, "Marked and Unmarked," 301–02; Brekhus, "A Sociology of the Unmarked," 35; Wårvik, "What Is Foregrounded in Narratives?" 105.
85. Asia Friedman, "Toward a Sociology of Perception: Sight, Sex, and Gender," *Cultural Sociology* 5 (2011): 197; Asia Friedman, *Blind to Sameness: Sexpectations and the Social Construction of Male and Female Bodies* (Chicago: University of Chicago Press, 2013), 110.

Chapter 3

1. Carl Purcell, *Your Artist's Brain* (Cincinnati, OH: North Light Books, 2010), 61.
2. See, for example, Calvin F. Nodine et al., "Searching for Nina," in John W. Senders et al. (eds.), *Eye Movements and the Higher Psychological Functions* (Hillsdale, NJ: Lawrence Erlbaum Associates, 1978), 242; Calvin F. Nodine et al., "Eye Movements During Visual Search for Artistically Embedded Targets," *Bulletin of the Psychonomic Society* 13 (1979): 371–74; Calvin F. Nodine and Harold L. Kundel, "Using Eye Movements to Study Visual Search and to Improve Tumor Detection," *RadioGraphics* 7 (1987): 1241–50.
3. See also Nodine and Kundel, "Using Eye Movements to Study Visual Search and to Improve Tumor Detection," 1241; Richard D. Zakia, *Perception and Imaging*, 2nd ed. (Boston: Focal Press, 2002), 6.

4. See also Richard Jung, "Correlation of Bioelectrical and Autonomic Phenomena with Alterations of Consciousness and Arousal in Man," in J. F. Delafresnaye (ed.), *Brain Mechanisms and Consciousness* (Springfield, IL: Charles C. Thomas, 1954), 311.
5. Joseph Jastrow, *The Subconscious* (Boston: Houghton, Mifflin & Co., 1906), 60.
6. Ulric Neisser, "Visual Search," *Scientific American* 210 (June 1964): 94.
7. Thomas H. Davenport and John C. Beck, *The Attention Economy: Understanding the New Currency of Business* (Boston: Harvard Business School Press, 2001), 60.
8. Gary A. Fine, *Morel Tales: The Culture of Mushrooming* (Cambridge, MA: Harvard University Press, 1998), 102.
9. See, for example, Charles Derber, *The Pursuit of Attention: Power and Ego in Everyday Life*, 2nd ed. (New York: Oxford University Press, 2000).
10. Charles Goodwin, "Professional Vision," *American Anthropologist* 96 (1994): 610.
11. Ibid., 628, 610. Emphasis added.
12. See also Pierre L. Divenyi and Ira J. Hirsh, "Some Figural Properties of Auditory Patterns," *Journal of the Acoustical Society of America* 64 (1978): 1383; Albert S. Bregman, *Auditory Scene Analysis: The Perceptual Organization of Sound* (Cambridge, MA: MIT Press, 1990), 466.
13. See, for example, Bregman, *Auditory Scene Analysis*, 398, 497; Andrew H. Gregory, "Listening to Polyphonic Music," *Psychology of Music* 18 (1990): 169; David Huron, "Tonal Consonance versus Tonal Fusion in Polyphonic Sonorities," *Music Perception* 9 (1991): 135–54; E. Bigand et al., "Divided Attention in Music," *International Journal of Psychology* 35 (2000): 271. See also David Huron, "Voice Denumerability in Polyphonic Music of Homogeneous Timbres," *Music Perception* 6 (1989): 369.
14. See also Robert Ornstein, *Meditation and Modern Psychology* (Los Altos, CA: Malor Books, 2008 [1971]), 30.
15. Walter Cohen, "Spatial and Textural Characteristics of the Ganzfeld," *American Journal of Psychology* 70 (1957): 407.
16. See, for example, Cohen, "Spatial and Textural Characteristics of the Ganzfeld"; Walter Cohen, "Form Recognition, Spatial Orientation, Perception of Movement in the Uniform Visual Field,"

in Ailene Morris and E. Porter Horne (eds.), *Visual Search Techniques* (Washington, DC: Natural Academy of Sciences, 1960), 119; Magdalen D. Vernon, *The Psychology of Perception* (Harmondsworth, UK: Penguin, 1962), 45; Lloyd L. Avant, "Vision in the Ganzfeld," *Psychological Bulletin* 64 (1965): 246–58; Ornstein, *Meditation and Modern Psychology*, 33; Jirí Wackermann et al., "Ganzfeld-Induced Hallucinatory Experience, Its Phenomenology and Electrophysiology," *Cortex* 44 (2008): 1364–78.
17. See also Bregman, *Auditory Scene Analysis*, 469, 493.
18. See, for example, Carl M. Francolini and Howard E. Egeth, "Perceptual Selectivity Is Task Dependent: The Pop-Out Effect Poops Out," *Perception and Psychophysics* 25 (1979): 99–110.
19. Purcell, *Your Artist's Brain*, 61.
20. Katharine Q. Seelye et al., "F.B.I. Posts Images of Pair Suspected in Boston Attack," *New York Times*, April 19, 2013, A20, http://www.nytimes.com/2013/04/19/us/fbi-releases-video-of-boston-bombing-suspects.html?_r=0&pagewanted=print (accessed on August 7, 2013).
21. Edgar A. Poe, "The Purloined Letter," in *A Collection of Stories* (New York: Tom Doherty Associates, 1988 [1844]), 188–208.
22. See also Harold Garfinkel, "Studies of the Routine Grounds of Everyday Activities," in *Studies in Ethnomethodology* (Englewood Cliffs, NJ: Prentice-Hall, 1967 [1964]), 35–75; Rolf Kjolseth, "Making Sense: Natural Language and Shared Knowledge in Understanding," in Joshua A. Fishman (ed.), *Advances in the Sociology of Language, Volume II: Selected Studies and Applications* (The Hague, The Netherlands: Mouton, 1972), 60; Alfred Schutz and Thomas Luckmann, *The Structures of the Life-World* (Evanston, IL: Northwestern University Press, 1973), 3–4, 8; Robert Stalnaker, "Presuppositions," *Journal of Philosophical Logic* 2 (1973): 447; Douglas R. Hofstadter, "Changes in Default Words and Images, Engendered by Rising Consciousness," in *Metamagical Themas: Questing for the Essence of Mind and Pattern* (New York: Basic Books, 1985 [1982]), 137; Diane Ackerman, *A Natural History of the Senses* (New York: Random House, 1990), 235; Ross Chambers, "The Unexamined," in Mike Hill (ed.), *Whiteness: A Critical Reader* (New York: New York University Press, 1997), 189, 193–94; Wayne H. Brekhus, "A Sociology of the Unmarked: Redirecting Our Focus," *Sociological Theory* 16 (1998): 35.

23. Ludwig Wittgenstein, *Philosophical Investigations*, 4th ed. (Malden, MA: Wiley-Blackwell, 2009 [1953]), 56.
24. Olga Bogdashina, *Sensory Perceptual Issues in Autism and Asperger Syndrome: Different Sensory Experiences—Different Perceptual Worlds* (London: Jessica Kingsley Publishers, 2003), 36.
25. See also William James, *The Principles of Psychology* (Cambridge, MA: Harvard University Press, 1983 [1890]), 430.
26. See Brekhus, "A Sociology of the Unmarked."
27. John Urry, *The Tourist Gaze: Leisure and Travel in Contemporary Societies* (London: SAGE Publications, 1990), 3.
28. Erving Goffman, "Alienation from Interaction," in *Interaction Ritual: Essays on Face-to-Face Behavior* (Garden City, NY: Anchor Books, 1967 [1957]), 131; Erving Goffman, "Fun in Games," in *Encounters: Two Studies in the Sociology of Interaction* (Indianapolis, IN: Bobbs-Merrill, 1961), 63, 65–66; Erving Goffman, *Behavior in Public Places: Notes on the Social Organization of Gatherings* (New York: Free Press, 1963), 91; Erving Goffman, "Footing," in *Forms of Talk* (Philadelphia: University of Pennsylvania Press, 1981 [1979]), 131–32.
29. Eviatar Zerubavel, *Social Mindscapes: An Invitation to Cognitive Sociology* (Cambridge, MA: Harvard University Press, 1997), 39. See also Erving Goffman, *Relations in Public: Microstudies of the Public Order* (New York: Harper & Row, 1971), 307.
30. See, for example, Ralph Ellison, *Invisible Man* (New York: Random House, 1952).
31. Erving Goffman, *The Presentation of Self in Everyday Life* (Garden City, NY: Doubleday Anchor, 1959), 151–52.
32. Derber, *The Pursuit of Attention*, 15.
33. Goffman, "Footing," 132. See also Erving Goffman, *Frame Analysis: An Essay on the Organization of Experience* (New York: Harper & Row, 1974), 207.
34. Derber, *The Pursuit of Attention*, 35–55, 69–70. See also Andrew Duncan, *Centre and Periphery in Modern British Poetry* (Liverpool, UK: Liverpool University Press, 2005), 32.
35. Katharine Q. Seelye and Ian Lovett, "After Attack, Suspects Returned to Routines, Raising No Suspicions," *New York Times*, April 27, 2013, A12.
36. Wolfgang Sofsky, *The Order of Terror: The Concentration Camp* (Princeton, NJ: Princeton University Press, 1996 [1993]), 216.

37. Wayne H. Brekhus, *Peacocks, Chameleons, Centaurs: Gay Suburbia and the Grammar of Social Identity* (Chicago: University of Chicago Press, 2003), 29, 58–64. See also William R. Force, "The Code of Harry: Performing Normativity in Dexter," *Crime, Media, Culture* 6 (2010): 335–36, 340.
38. Robert Baden-Powell, *My Adventures as a Spy* (Cirencester, UK: The Echo Library, 2005 [1915]), 26–27.
39. John Hotchkiss, "Children and Conduct in a Ladino Community in Chiapas, Mexico," *American Anthropologist* 69 (1967): 711–18; Barbara Rogoff, *Apprenticeship in Thinking: Cognitive Development in Social Context* (New York: Oxford University Press, 1990), 124–26.
40. Purcell, *Your Artist's Brain*, 61.
41. Colin Watson, *Lonelyheart 4122: A Flaxborough Novel* (Bath, UK: Chivers Press, 1993), 33.
42. See also Robert Gottsdanker, "The Relation between the Nature of the Search Situation and the Effectiveness of Alternative Strategies of Search," in Ailene Morris and E. Porter Horne (eds.), *Visual Search Techniques* (Washington, DC: Natural Academy of Sciences, 1960), 181; John Duncan and Glyn W. Humphreys, "Visual Search and Stimulus Similarity," *Psychological Review* 96 (1989): 434, 442.
43. Nodine et al., "Eye Movements During Visual Search for Artistically Embedded Targets," 371.
44. Geoffrey Barkas, in Seymour Reit, *Masquerade: The Amazing Camouflage Deceptions of World War II* (New York: Hawthorn Books, 1978), 92.
45. See also Gottsdanker, "The Relation between the Nature of the Search Situation and the Effectiveness of Alternative Strategies of Search," 181; Tom Troscianko et al., "Camouflage and Visual Perception," *Philosophical Transactions of the Royal Society B: Biological Sciences* 364 (2009): 456; Graeme D. Ruxton, "Non-Visual Crypsis: A Review of the Empirical Evidence of Camouflage to Senses Other than Vision," *Philosophical Transactions of the Royal Society B: Biological Sciences* 364 (2009): 549–50, 552; Martin Stevens and Sami Merilaita, "Animal Camouflage: Current Issues and New Perspectives," *Philosophical Transactions of the Royal Society B: Biological Sciences* 364 (2009): 425.

46. See also Wolfgang Köhler, *Gestalt Psychology: An Introduction to New Concepts in Modern Psychology* (New York: Mentor Books, 1947), 84; Zakia, *Perception and Imaging*, 121–22.
47. See also Daniel C. Dennett, *Consciousness Explained* (Boston: Little, Brown & Co., 1991), 335.
48. See also Innes C. Cuthill et al., "Disruptive Coloration and Background Pattern Matching," *Nature* 434 (March 3, 2005): 72; Hanna R. Shell, *Hide and Seek: Camouflage, Photography, and the Media of Reconnaissance* (New York: Zone Books, 2012), 47.
49. John A. Endler, "A Predator's View of Animal Color Patterns." *Evolutionary Biology* 11 (1978): 319.
50. See, for example, Gerald Thayer and Abbott H. Thayer, *Concealing-Coloration in the Animal Kingdom: An Exposition of the Laws of Disguise through Color and Pattern* (New York: Macmillan, 1909), 10; Angeline M. Keen, "Protective Coloration in the Light of Gestalt Theory," *Journal of General Psychology* 6 (1932): 200; Wolfgang Metzger, *Laws of Seeing* (Cambridge, MA: MIT Press, 2006 [1936]), 64–65; Roy R. Behrens, *False Colors: Art, Design, and Modern Camouflage* (Dysart, IA: Bobolink Books, 2002), 117; Zakia, *Perception and Imaging*, 17; Roy R. Behrens, *Camoupedia: A Compendium of Research on Art, Architecture and Camouflage* (Dysart, IA: Bobolink Books, 2009), 59; Innes C. Cuthill and Tom S. Troscianko, "Animal Camouflage: Biology Meets Psychology, Computer Science, and Art," in C. A. Brebbia et al. (eds.), *Colour in Art, Design, and Nature* (Ashurst, UK: WIT Press, 2011), 10.
51. Adolf Portmann, *Animal Camouflage* (Ann Arbor: University of Michigan Press, 1959), 9.
52. Erasmus Darwin, *Zoonomia, or the Laws of Organic Life* (London: J. Johnson, 1794), Vol. 1, 577–78.
53. Alfred R. Wallace, *Darwinism: An Exposition of the Theory of Natural Selection with Some of Its Applications* (London: Macmillan, 1891), 190. See also Metzger, *Laws of Seeing*, 64.
54. Edward B. Poulton, *The Colours of Animals* (New York: D. Appleton, 1890), 24. Emphasis added.
55. Thayer and Thayer, *Concealing-Coloration in the Animal Kingdom*, 13. See also 3.
56. Abbott H. Thayer, "Disruptive Camouflage," in Roy R. Behrens (ed.), *Ship Shape: A Dazzle Camouflage Sourcebook* (Dysart, IA:

Bobolink Books, 2012 [1918]), 41. Emphasis added. See also Thayer and Thayer, *Concealing-Coloration in the Animal Kingdom*, 5–7.
57. Thayer and Thayer, *Concealing-Coloration in the Animal Kingdom*, 12.
58. Frank E. Beddard, *Animal Coloration: An Account of the Principal Facts and Theories Relating to the Colours and Markings of Animals* (London: Swan Sonnenschein, 1895), 85.
59. Thayer and Thayer, *Concealing-Coloration in the Animal Kingdom*, 10.
60. Henry S. Williams, "Natural Selection and Ship Camouflage," in Roy R. Behrens (ed.), *Ship Shape: A Dazzle Camouflage Sourcebook* (Dysart, IA: Bobolink Books, 2012 [1919]), 86.
61. Thayer, "Disruptive Camouflage," 41.
62. Ibid.
63. See also Shell, *Hide and Seek*, 46–53.
64. Thayer, "Disruptive Camouflage," 42.
65. Darwin, *Zoonomia*, 577; Poulton, *The Colours of Animals*, 24. See also Keen, "Protective Coloration in the Light of Gestalt Theory," 202.
66. Wallace, *Darwinism*, 190. See also 200.
67. Thayer and Thayer, *Concealing-Coloration in the Animal Kingdom*, 7.
68. Stephen Kern, *The Culture of Time and Space 1880–1918* (Cambridge, MA: Harvard University Press, 1983), 302; Behrens, *False Colors*, 54; Hardy Blechman, *Disruptive Pattern Manual: An Encyclopedia of Camouflage* (Buffalo, NY: Firefly Books, 2004), 120–23; Tim Newark, *Camouflage* (London: Thames and Hudson, 2007), 40–46; Behrens, *Camoupedia*, 59.
69. Kern, *The Culture of Time and Space 1880–1918*, 302; Stephen Kern, "Cubism, Camouflage, Silence, and Democracy: A Phenomenological Approach," in Roger Friedland and Deirdre Boden (eds.), *NowHere: Space, Time, and Modernity* (Berkeley: University of California Press, 1994), 166; Blechman, *Disruptive Pattern Manual*, 123.
70. Newark, *Camouflage*, 46.
71. Behrens, *Camoupedia*, 19–21. See also Herbert G. Wells, *Italy, France and Britain at War* (New York: Macmillan, 1917), 111; Paul K. Saint-Amour, "Modernist Reconnaissance," *Modernism/Modernity* 10 (2003): 349–80; Newark, *Camouflage*, 54.

72. Kern, *The Culture of Time and Space 1880–1918*, 303; Blechman, *Disruptive Pattern Manual*, 42, 274–75.
73. Kern, *The Culture of Time and Space 1880–1918*, 303; Kern, "Cubism, Camouflage, Silence, and Democracy," 166; Behrens, *False Colors*, 66–67; Blechman, *Disruptive Pattern Manual*, 128–29, 244–45; Newark, *Camouflage*, 54.
74. Blechman, *Disruptive Pattern Manual*, 132–33.
75. See, for example, http://www.veruschka.net/ (accessed on September 14, 2013); http://trendland.com/rachel-perry-welty-lost-my-life/ (accessed on December 25, 2012); http://www.desireepalmen.nl/images.php (accessed on September 14, 2013); http://laurentlagamba.free.fr/ (accessed on September 14, 2013).
76. See also http://www.moillusions.com/tag/liu-bolin (accessed on February 3, 2012).
77. See also Blechman, *Disruptive Pattern Manual*, 306, 309, 314.
78. See, for example, Ruxton, "Non-Visual Crypsis," 552.
79. See, for example, Steve Chapman, *365 Things Every Hunter Should Know* (Eugene, OR: Harvest House Publishers, 2008), 133–35, 138; Bill Vaznis, *500 Deer Hunting Tips: Strategies, Techniques, and Methods* (Minneapolis, MN: Creative Publishing International, 2008), 39–40; Michael Collins, "Smell Invisible: Being a Scent Free Hunter," *Cincinnati Hunting Examiner*, September 12, 2011, http://www.examiner.com/hunting-in-cincinnati/smell-invisible-being-a-scent-free-hunter (accessed on December 16, 2011). See also Lindsay Wood, "How to Remove Scent from Hunting Gear," http://www.ehow.com/how_5663627_remove-scent-hunting-gear.html (accessed on December 16, 2011).
80. http://www.flickr.com/photos/23891490@N00/62344065/ (accessed on August 2, 2013).
81. Jacqueline L. Tobin and Raymond G. Dobard, *Hidden in Plain View: A Secret Story of Quilts and the Underground Railroad* (New York: Anchor Books, 2000 [1999]).
82. William C. Kashatus, *Just Over the Line: Chester County and the Underground Railroad* (West Chester, PA: Chester County Historical Society, 2002), 18–19.
83. Sean Monahan, "Camouflaged Communication: Hiding a Connection between Signifier and Signified" (unpublished paper, Rutgers University, New Brunswick, NJ, 2002).

84. http://www.military.com/video/operations-and-strategy/vietnam-war/pow-blinks-torture-in-morse-code/1381254901001/ (accessed on September 17, 2013); Jeremiah A. Denton and Edwin H. Brandt, *When Hell Was in Session* (Traditional Press, 1982), 70.
85. Richard Huff and Brian Kates, "TV to Screen Bin Laden Tapes: White House Warns They May Contain Coded Messages," *New York Daily News*, October 11, 2001, http://articles.nydailynews.com/2001-10-11/news/18369890_1_laden-bin-messages (accessed on December 16, 2011).
86. Stefan A. Pluta, "Affidavit in Support of Criminal Complaint, Arrest Warrant and Search Warrants against Robert Philip Hanssen," February 2001, http://www.fbi.gov/about-us/history/famous-cases/robert-hanssen/affidavit.pdf (accessed on September 17, 2013), 27–28.
87. Saint-Amour, "Modernist Reconnaissance," 370.
88. Martin Stevens and Sami Merilaita, "Defining Disruptive Coloration and Distinguishing Its Functions," *Philosophical Transactions of the Royal Society B: Biological Sciences* 364 (2009): 481, 483. See also Stevens and Merilaita, "Animal Camouflage," 424.
89. Thayer and Thayer, *Concealing-Coloration in the Animal Kingdom*, 98. See also 77–79.
90. Metzger, *Laws of Seeing*, 73.
91. Sami Merilaita and Johan Lind, "Background-Matching and Disruptive Coloration, and the Evolution of Cryptic Coloration," *Proceedings of the Royal Society B: Biological Sciences* 272 (2005): 665. See also Jeremy M. Wolfe et al., "Segmentation of Objects from Backgrounds in Visual Search Tasks," *Vision Research* 42 (2002): 2985–86; Cuthill and Troscianko, "Animal Camouflage," 16, 18–19.
92. Troscianko et al., "Camouflage and Visual Perception," 451.
93. Merilaita and Lind, "Background-Matching and Disruptive Coloration, and the Evolution of Cryptic Coloration," 665.
94. Blechman, *Disruptive Pattern Manual*, 27.
95. See also ibid., 154–57; Newark, *Camouflage*, 51.
96. See, for example, Behrens, *Camoupedia*, 40; Blechman, *Disruptive Pattern Manual*, 250. See also Wells, *Italy, France and Britain at War*, 112–13; Reit, *Masquerade*, 93.

97. Kurt Koffka, *Principles of Gestalt Psychology* (New York: Harcourt, Brace & World, 1935), 77. Emphasis added.
98. See also Matthew Luckiesh, *Visual Illusions: Their Causes, Characteristics, and Applications* (New York: D. Van Nostrand & Constable, 1922), 231; Behrens, *False Colors*, 91, 99; Blechman, *Disruptive Pattern Manual*, 164–70; Behrens, *Camoupedia*, 10, 380–82; Roy R. Behrens (ed.), *Ship Shape: A Dazzle Camouflage Sourcebook* (Dysart, IA: Bobolink Books, 2012).
99. See Lionel Casson, *Ships and Seamanship in the Ancient World* (Baltimore, MD: Johns Hopkins University Press, 1995 [1971]), 235n47.
100. Köhler, *Gestalt Psychology*, 92–93. Emphasis added.
101. See also ibid., 92; Kate Plaisted et al., "Enhanced Visual Search for a Conjunctive Target in Autism: A Research Note," *Journal of Child Psychology and Psychiatry* 39 (1998): 778; Troscianko et al., "Camouflage and Visual Perception," 458.
102. See also Duncan and Humphreys, "Visual Search and Stimulus Similarity," 451.
103. Harold L. Kundel and Calvin F. Nodine, "Studies of Eye Movements and Visual Search in Radiology," in John W. Senders et al. (eds.), *Eye Movements and the Higher Psychological Functions* (Hillsdale, NJ: Lawrence Erlbaum Associates, 1978), 324; Duncan and Humphreys, "Visual Search and Stimulus Similarity," 434, 442. See also Lila Ghent, "Perception of Overlapping Figures by Children of Different Ages," *American Journal of Psychology* 69 (1956): 584–85; Herman A. Witkin and John W. Berry, "Psychological Differentiation in Cross-Cultural Perspective," *Journal of Cross-Cultural Psychology* 6 (1975): 7.
104. Paul T. Sowden et al., "Perceptual Learning of the Detection of Features in X-Ray Images: A Functional Role for Improvements in Adults' Visual Sensitivity?" *Journal of Experimental Psychology: Human Perception and Performance* 26 (2000): 379. See also Nodine and Kundel, "Using Eye Movements to Study Visual Search and to Improve Tumor Detection," 1242–43; Elizabeth A. Krupinski, "Visual Scanning Patterns of Radiologists Searching Mammograms," *Academic Radiology* 3 (1996): 137.
105. Blechman, *Disruptive Pattern Manual*, 77.
106. Wylie Sypher, *Rococo to Cubism in Art and Literature* (New York: Random House, 1960), 270. Emphasis added.

107. Behrens, *False Colors*, 68–69. See also Kern, *The Culture of Time and Space 1880–1918*, 302–03; Saint-Amour, "Modernist Reconnaissance," 349–55.
108. See Kurt Gottschaldt, "The Influence of Past Experience on the Perception of Figures," in M. D. Vernon (ed.), *Experiments in Visual Perception: Selected Readings* (Harmondsworth, UK: Penguin Books, 1966 [1926]), 32–33. See also William Linden, "Practicing of Meditation by School Children and Their Levels of Field Dependence-Independence, Test Anxiety, and Reading Achievement," *Journal of Consulting and Clinical Psychology* 41 (1973): 139.
109. See also Nodine et al., "Searching for Nina."
110. See, for example, Witkin and Berry, "Psychological Differentiation in Cross-Cultural Perspective," 7; Divenyi and Hirsh, "Some Figural Properties of Auditory Patterns," 1370.
111. Dennett, *Consciousness Explained*, 334.
112. http://en.wikipedia.org/wiki/Hidden_Mickey (accessed on July 24, 2013).
113. See also Behrens (ed.), *Ship Shape*, 188–89; Köhler, *Gestalt Psychology*, 92; Nodine et al., "Searching for Nina," 241; Nodine et al., "Eye Movements During Visual Search for Artistically Embedded Targets," 371; Roy R. Behrens, "Revisiting Gottschaldt: Embedded Figures in Art, Architecture, and Design," *Gestalt Theory: Journal of the GTA* 22, no. 2 (2000): 97–106; Behrens, *Camoupedia*, 131–32.
114. See, for example, Sarah L. Schuette, *Pets All Around: A Spot-It Challenge* (North Mankato, MN: Capstone Press, 2013); Sarah L. Schuette, *Sports Zone: A Spot-It Challenge* (North Mankato, MN: Capstone Press, 2013); Martin Handford, *Where's Waldo?* (Somerville, MA: Candlewick Press, 2007 [1987]).
115. Joseph Rosenbloom, *Doctor Knock-Knock's Official Knock-Knock Dictionary* (New York: Sterling, 1976), 53. See also 58; Eviatar Zerubavel, *The Fine Line: Making Distinctions in Everyday Life* (Chicago: University of Chicago Press, 1993 [1991]), 92–93.
116. See also Daniel Memmert and Philip Furley, "'I Spy with My Little Eye!': Breadth of Attention, Inattentional Blindness, and Tactical Decision Making in Team Sports," *Journal of Sport and Exercise Psychology* 29 (2007): 365.

117. See also Bridgette Burch, "The Misdirection in Attention" (unpublished paper, Rutgers University, New Brunswick, NJ, 2012).
118. See also Rosemarie Garland-Thomson, *Staring: How We Look* (New York: Oxford University Press, 2009), 15.
119. François Truffaut, *Hitchcock*, rev. ed. (New York: Touchstone Books, 1985 [1983]), 269. Emphasis added.
120. Reit, *Masquerade*, 19. Emphasis added. See also Warren E. Steinkraus, "The Art of Conjuring," *Journal of Aesthetic Education* 13, no. 4 (October 1979): 23; Eddie Joseph, *How to Pick Pockets for Fun and Profit: A Magician's Guide to Pickpocket Magic* (Colorado Springs, CO: Piccadilly Books, 1992), 13; Nathaniel Schiffman, *Abracadabra: Secret Methods Magicians and Others Use to Deceive Their Audience* (Amherst, NY: Prometheus Books, 1997), 17–18, 369; Behrens, *False Colors*, 118, 160, 162–65; William R. Freudenburg and Margarita Alario, "Weapons of Mass Distraction: Magicianship, Misdirection, and the Dark Side of Legitimation," *Sociological Forum* 22 (2007): 148; Gustav Kuhn and John M. Findlay, "Misdirection, Attention, and Awareness: Inattentional Blindness Reveals Temporal Relationship between Eye Movements and Visual Awareness," *Quarterly Journal of Experimental Psychology* 63 (2010): 143.
121. Schiffman, *Abracadabra*, 18.
122. Gustav Kuhn et al., "Towards a Science of Magic," *Trends in Cognitive Sciences* 12 (2008): 349. See also Peter Lamont and Richard Wiseman, *Magic in Theory: An Introduction to the Theoretical and Psychological Elements of Conjuring* (Hatfield, UK: University of Hertfordshire Press, 1999), 39, 46.
123. Schiffman, *Abracadabra*, 18.
124. Ibid., 405.
125. Hugh B. Cott, in Behrens, *False Colors*, 163. See also 118; Joseph, *How to Pick Pockets for Fun and Profit*, 12, 19; Stephen L. Macknik et al., "Attention and Awareness in Stage Magic: Turning Tricks into Research," *Nature Reviews Neuroscience* 9 (2008): 875; Behrens, *Camoupedia*, 256.
126. Joseph, *How to Pick Pockets for Fun and Profit*, 10; Schiffman, *Abracadabra*, 367–68.
127. Sun Tzu, *The Art of War* (Oxford: Oxford University Press, 1963 [sixth century BC]), 51.

128. Reit, *Masquerade*, 22–33.
129. Ibid., 33–43.
130. See also Freudenburg and Alario, "Weapons of Mass Distraction."
131. Joel de la Cruz, http://www.usingenglish.com/reference/idioms/wag+the+dog.html (accessed on January 14, 2012). Emphasis added.
132. Ibid. Emphasis added.

Chapter 4

1. Asia Friedman, *Blind to Sameness: Sexpectations and the Social Construction of Male and Female Bodies* (Chicago: University of Chicago Press, 2013), 23.
2. See, for example, Bruno G. Breitmeyer, *Blindspots: The Many Ways We Cannot See* (New York: Oxford University Press, 2010), 13.
3. James J. Gibson, *The Senses Considered as Perceptual Systems* (Boston: Houghton Mifflin, 1966), 174.
4. Breitmeyer, *Blindspots*, 11–12. See also Temple Grandin and Catherine Johnson, *Animals in Translation: Using the Mysteries of Autism to Decode Animal Behavior* (Orlando, FL: Harvest Books, 2006 [2005]), 40–41.
5. See also Richard D. Zakia, *Perception and Imaging*, 2nd ed. (Boston: Focal Press, 2002), 3.
6. Herbert J. Schlesinger, "Cognitive Attitudes in Relation to Susceptibility to Interference," *Journal of Personality* 22 (1954): 356.
7. Joseph Kasof, "Creativity and Breadth of Attention," *Creativity Research Journal* 10 (1997): 303.
8. Herman A. Witkin et al., *Psychological Differentiation: Studies of Development* (New York: John Wiley and Sons, 1962), 1–2, 57–58; Herman A. Witkin and John W. Berry, "Psychological Differentiation in Cross-Cultural Perspective," *Journal of Cross-Cultural Psychology* 6 (1975): 6.
9. See, for example, Witkin et al., *Psychological Differentiation*, 57; Eleanor Maccoby, "Sex Differences in Intellectual Functioning," in *The Development of Sex Differences* (Palo Alto, CA: Stanford University Press, 1966), 26–27; Eviatar Zerubavel, *The*

Fine Line: Making Distinctions in Everyday Life (Chicago: University of Chicago Press, 1993 [1991]), 50, 116. See also Richard E. Nisbett, *The Geography of Thought: How Asians and Westerners Think Differently . . . and Why* (New York: Free Press, 2003).

10. See, for example, Ami Klin et al., "Visual Fixation Patterns During Viewing of Naturalistic Social Situations as Predictors of Social Competence in Individuals with Autism," *Archives of General Psychiatry* 59 (2002): 809–16; Warren Jones and Ami Klin, "Attention to Eyes Is Present but in Decline in 2-6-Month-Old Infants Later Diagnosed with Autism," *Nature* 12715 (November 6, 2013).

11. See, for example, Richard D. Halley, "Distractability of Males and Females in Competing Aural Message Situations: A Research Note," *Human Communication Research* 2 (1975): 79–82.

12. See, for example, Herman A. Witkin et al., *Personality through Perception: An Experimental and Clinical Study* (New York: Harper & Brothers, 1954), 151, 159, 170, 482; James Bieri et al., "Sex Differences in Perceptual Behavior," *Journal of Personality* 26 (1958): 1–12; Witkin et al., *Psychological Differentiation*, 214–21; Martin J. Doherty et al., "The Context-Sensitivity of Visual Size Perception Varies across Cultures," *Perception* 37 (2008): 1426–33.

13. See, for example, Anne Moir and David Jessel, *Brain Sex: The Real Difference Between Men and Women* (New York: Lyle Stuart, 1991), 18; Barbara Pease and Allan Pease, *Why Men Don't Listen and Women Can't Read Maps: How We're Different and What to Do About It* (New York: Broadway Books, 2001 [2000]), 21–22.

14. Pease and Pease, *Why Men Don't Listen and Women Can't Read Maps*, 21. See also Mary C. Bateson, *Peripheral Visions: Learning along the Way* (New York: HarperCollins, 1994), 97.

15. See also Julia A. Sherman, "Problem of Sex Differences in Space Perception and Aspects of Intellectual Functioning," *Psychological Review* 74 (1967): 295–98.

16. See, for example, John W. Berry, "Temne and Eskimo Perceptual Skills," *International Journal of Psychology* 1 (1966): 225–28.

17. See also Eviatar Zerubavel, *Social Mindscapes: An Invitation to Cognitive Sociology* (Cambridge, MA: Harvard University Press, 1997), 39.

18. Peter Singer, *The Expanding Circle: Ethics and Sociobiology* (New York: Farrar, Straus & Giroux, 1981).
19. See, however, Zerubavel, *The Fine Line*, 104–05.
20. See also Murray S. Davis, *Smut: Erotic Reality/Obscene Ideology* (Chicago: University of Chicago Press, 1983), 27–30, 40–41, 133–39, 144–50, 156–59; Zerubavel, *Social Mindscapes*, 52.
21. See also Zerubavel, *Social Mindscapes*, 33–34, 46–47.
22. See also ibid., 33.
23. Frederic C. Bartlett, *Remembering: A Study in Experimental and Social Psychology* (Cambridge: Cambridge University Press, 1964 [1932]), 255. See also 253–54.
24. See, for example, Ian Watson, "How Tour Operators and Travel Guidebooks Select Destinations" (lecture given at the "Practicing Nature-Based Tourism" Conference, Reykjavik Art Museum, February 2011). See also Samuel C. Heilman, *A Walker in Jerusalem* (New York: Summit Books, 1986), 86, 100, 107, 110; John Urry, *The Tourist Gaze: Leisure and Travel in Contemporary Societies* (London: SAGE Publications, 1990), 66, 138–40.
25. Zerubavel, *Social Mindscapes*, 50; Kari M. Norgaard, *Living in Denial: Climate Change, Emotions, and Everyday Life* (Cambridge, MA: MIT Press, 2011), 112. See also Erving Goffman, "Fun in Games," in *Encounters: Two Studies in the Sociology of Interaction* (Indianapolis, IN: Bobbs-Merrill, 1961), 19–22 on "rules of irrelevance."
26. Friedman, *Blind to Sameness*, 23.
27. Ibid., 2.
28. Ibid., 29. See also 13.
29. Zerubavel, *Social Mindscapes*, 35–52.
30. I borrow the term "socio-attentional" from Stephanie E. Alves, "Blind Trusting the Blind: The Attentional Norms of Social Trust" (unpublished manuscript, Rutgers University, New Brunswick, NJ, 2013).
31. Berry, "Temne and Eskimo Perceptual Skills," 135. See also Daniel Kahneman, *Attention and Effort* (Englewood Cliffs, NJ: Prentice-Hall, 1973), 78; Leonard Shlain, *The Alphabet versus the Goddess: The Conflict between Word and Image* (New York: Penguin Arkana, 1998), 25–26.
32. Berry, "Temne and Eskimo Perceptual Skills," 212.

33. See also John W. Berry, "Ecological and Cultural Factors in Spatial Perceptual Development," *Canadian Journal of Behavioural Science* 3 (1971): 324–36; Witkin and Berry, "Psychological Differentiation in Cross-Cultural Perspective," 59–62.
34. See, for example, Ulrich Kühnen et al., "Cross-Cultural Variations in Identifying Embedded Figures: Comparisons from the United States, Germany, Russia, and Malaysia," *Journal of Cross-Cultural Psychology* 32 (2001): 369–70. See also Robert Schwitzgebel, "The Performance of Dutch and Zulu Adults in Selected Perceptual Tasks," *Journal of Social Psychology* 57 (1962): 75–76.
35. See, for example, Elinor Ochs, *Culture and Language Development: Language Acquisition and Language Socialization in a Samoan Village* (New York: Cambridge University Press, 1988), 47; Barbara Rogoff et al., "Toddlers' Guided Participation with Their Caregivers in Cultural Activity," in Ellice A. Forman et al. (eds.), *Contexts for Learning: Sociocultural Dynamics in Children's Development* (New York: Oxford University Press, 1996), 244; Pablo Chavajay and Barbara Rogoff, "Cultural Variation in Management of Attention by Children and Their Caregivers," *Developmental Psychology* 35 (1999): 1079–90. See also Maricela Correa-Chávez et al., "Cultural Patterns in Attending to Two Events at Once," *Child Development* 76 (2005): 664–78.
36. Pliny the Elder. *The Natural History of Pliny*, vol. 6 (London: Henry G. Bohn, 1857 [circa AD 77]), 283 (the 43rd chapter of the 35th book of his *Natural History*).
37. See also Janice Haaken, "Field Dependence Research: A Historical Analysis of a Psychological Construct," *Signs* 13 (1988): 311–30.
38. Ayse K. Uskul et al., "Ecocultural Basis of Cognition: Farmers and Fishermen Are More Holistic than Herders," *Proceedings of the National Academy of Sciences of the United States of America* 105 (June 24, 2008): 8554.
39. Hazel R. Markus and Shinobu Kitayama, "Culture and the Self: Implications for Cognition, Emotion, and Motivation," *Psychological Review* 98 (1991): 246.
40. Uskul et al., "Ecocultural Basis of Cognition," 8552.
41. Nisbett, *The Geography of Thought*. See also Edward T. Hall, *Beyond Culture* (Garden City, NY: Anchor Books, 1977 [1976]), 91; Kühnen et al., "Cross-Cultural Variations in Identifying Embedded Figures," 366–67.

42. Li-Jun Ji et al., "Culture, Control, and Perception of Relationships in the Environment," *Journal of Personality and Social Psychology* 78 (2000): 951–52; Takahiko Masuda and Richard E. Nisbett, "Attending Holistically versus Analytically: Comparing the Context Sensitivity of Japanese and Americans," *Journal of Personality and Social Psychology* 81 (2001): 929–33; Nisbett, *The Geography of Thought*, 82, 100; Takahiko Masuda et al., "Culture and Aesthetic Preference: Comparing the Attention to Context of East Asians and Americans," *Personality and Social Psychology Bulletin* 34 (2008): 1262.
43. Richard E. Nisbett and Yuri Miyamoto, "The Influence of Culture: Holistic vs. Analytic Perception," *Trends in Cognitive Sciences* 9 (2005): 467.
44. Shinobu Kitayama et al., "Perceiving an Object and Its Context in Different Cultures: A Cultural Look at New Look," *Psychological Science* 14 (2003): 201; Hannah F. Chua et al., "Cultural Variation in Eye Movements During Scene Perception," *Proceedings of the National Academy of Sciences of the United States of America* 102 (August 30, 2005): 12629; Aysecan Boduroglu et al., "Cultural Differences in Allocation of Attention in Visual Information Processing," *Journal of Cross-Cultural Psychology* 40 (2009): 349. See also Masuda and Nisbett, "Attending Holistically versus Analytically," 933; Angela H. Gutchess et al., "Cultural Differences in Neural Function Associated with Object Processing," *Cognitive, Affective, and Behavioral Neuroscience* 6 (2006): 105, 107; Takahiko Masuda and Richard E. Nisbett, "Culture and Change Blindness," *Cognitive Science* 30 (2006): 394; Masuda et al., "Culture and Aesthetic Preference," 1272.
45. See, for example, Masuda and Nisbett, "Attending Holistically versus Analytically," 929; Nisbett, *The Geography of Thought*, 92.
46. Masuda and Nisbett, "Attending Holistically versus Analytically," 927, 933. See also Kitayama et al., "Perceiving an Object and Its Context in Different Cultures," 201; Yuri Miyamoto et al., "Culture and the Physical Environment: Holistic versus Analytic Perceptual Affordances," *Psychological Science* 17 (2006): 118; Masuda et al., "Culture and Aesthetic Preference," 1261–62; Richard S. Lewis et al., "Culture and Context: East Asian American and European American Differences in P3

Event-Related Potentials and Self-Construal," *Personality and Social Psychology Bulletin* 34 (2008): 624.
47. Boduroglu et al., "Cultural Differences in Allocation of Attention in Visual Information Processing," 349, 351, 356; Hyekyung Park and Shinobu Kitayama, "Perceiving through Culture: The Socialized Attention Hypothesis," in Reginald B. Adams Jr. et al. (eds.), *The Science of Social Vision* (New York: Oxford University Press, 2011), 76.
48. Guy T. Buswell, *How People Look at Pictures: A Study of the Psychology of Perception in Art* (Chicago: University of Chicago Press, 1935), 9.
49. See, for example, Chua et al., "Cultural Variation in Eye Movements During Scene Perception"; Gutchess et al., "Cultural Differences in Neural Function Associated with Object Processing."
50. Chua et al., "Cultural Variation in Eye Movements During Scene Perception," 12631–32. See also Boduroglu et al., "Cultural Differences in Allocation of Attention in Visual Information Processing," 357.
51. Masuda et al., "Culture and Aesthetic Preference," 1262–63. See also 1264.
52. Ibid., 1263, 1267, 1271–72.
53. Peter L. Berger and Thomas Luckmann, *The Social Construction of Reality: A Treatise in the Sociology of Knowledge* (Garden City, NY: Anchor Books, 1967 [1966]), 77.
54. See also Charles Goodwin, "Professional Vision," *American Anthropologist* 96 (1994): 626.
55. Friedman, *Blind to Sameness*, 48.
56. Stefan Hirschauer, "The Manufacture of Bodies in Surgery," *Social Studies of Science* 21 (1991): 289. See also 288.
57. Ibid., 299. See also 297.
58. See also Doherty et al., "The Context-Sensitivity of Visual Size Perception Varies across Cultures."
59. See also Kristen Purcell, "In A League of Their Own: Mental Leveling and the Creation of Social Comparability in Sport," *Sociological Forum* 11 (1996): 421–33.
60. See also Jonathan Crary, *Suspensions of Perception: Attention, Spectacle, and Modern Culture* (Cambridge, MA: MIT Press, 1999), 1.

61. Urry, *The Tourist Gaze*, 129–30.
62. See, for example, John T. Lang, "Sound and the City: Noise in Restaurant Critics' Reviews" (paper presented at the annual meeting of the American Sociological Association, New York, August 2013).
63. See also Erving Goffman, "Alienation from Interaction," in *Interaction Ritual: Essays on Face-to-Face Behavior* (Garden City, NY: Anchor Books, 1967 [1957]), 113, 115; Erving Goffman, *Behavior in Public Places: Notes on the Social Organization of Gatherings* (New York: Free Press, 1963), 43–45.
64. For a list of smoking bans by country, see http://en.wikipedia.org/wiki/List_of_smoking_bans (accessed on February 4, 2013).
65. See also Christopher Stone, *Should Trees Have Standing? Toward Legal Rights for Natural Objects* (Los Altos, CA: William Kaufmann, 1974), 6–7.
66. Thomas S. Kuhn, *The Structure of Scientific Revolutions*, 2nd ed. (Chicago: University of Chicago Press, 1970 [1962]), 111.
67. See ibid., 116.
68. See Sigmund Freud, *The Psychopathology of Everyday Life* (New York: W. W. Norton, 1960 [1901]), 53–105.
69. See Edward T. Hall, *The Hidden Dimension* (Garden City, NY: Doubleday, 1966), 113–64.
70. Thomas Laqueur, *Making Sex: Body and Gender from the Greeks to Freud* (Cambridge, MA: Harvard University Press, 1990).
71. Friedman, *Blind to Sameness*, 49.
72. See, for example, Eviatar Zerubavel, "Language and Memory: 'Pre-Columbian' America and the Social Logic of Periodization," *Social Research* 65 (1998): 327–28; Eviatar Zerubavel, *Time Maps: Collective Memory and the Social Shape of the Past* (Chicago: University of Chicago Press, 2003), 27.
73. On that, see Eviatar Zerubavel, *Social Mindscapes*, 32.
74. Lisa Campion, "The Social Construction of Attention: Varying Patterns of Attending, Inattending, and Disattending within Healthy and Sick Communities" (unpublished manuscript, Rutgers University, New Brunswick, NJ, 2013).
75. Goffman, "Fun in Games," 19, 25.
76. Friedman, *Blind to Sameness*, 57.
77. See, for example, Emile Durkheim, *The Elementary Forms of Religious Life* (New York: Free Press, 1995 [1912]), 308–09; Rosemarie

Garland-Thomson, *Staring: How We Look* (New York: Oxford University Press, 2009), 5, 63–76.
78. Goffman, *Behavior in Public Places*, 84–87. See also Erving Goffman, *The Presentation of Self in Everyday Life* (Garden City, NY: Doubleday Anchor, 1959), 230; Goffman, "Fun in Games," 63.
79. Garland-Thomson, *Staring*, 35.
80. Michel Foucault, *The Birth of the Clinic: An Archaeology of Medical Perception* (New York: Vintage Books, 1975 [1963]), 107–23.
81. Joan P. Emerson, "Behavior in Private Places: Sustaining Definitions of Reality in Gynecological Examinations," in Hans-Peter Dreitzel (ed.), *Recent Sociology No. 2: Patterns of Communicative Behavior* (New York: Macmillan, 1970), 78, 83. See also 86.
82. See also Durkheim, *The Elementary Forms of Religious Life*, 308–09.
83. Judith Martin, *Miss Manners' Guide for the Turn-of-the-Millennium* (New York: Fireside, 1990), 100. See also 106–13; Goffman, *The Presentation of Self in Everyday Life*, 229.
84. Goffman, *The Presentation of Self in Everyday Life*, 231; Goffman, "Fun in Games," 55–56; Goffman, *Behavior in Public Places*, 84; Spencer E. Cahill et al., "Meanwhile Backstage: Public Bathrooms and the Interaction Order," *Urban Life* 14 (1985): 37; Douglas Mason-Schrock, "Transsexuals' Narrative Construction of the 'True Self,'" *Social Psychology Quarterly* 59 (1996): 189.
85. In François Truffaut's film *Stolen Kisses*.
86. Erving Goffman, "On Face Work: An Analysis of Ritual Elements in Social Interaction," in *Interaction Ritual: Essays on Face-to-Face Behavior* (Garden City, NY: Anchor Books, 1967 [1955]), 18.
87. Erving Goffman, *Relations in Public: Microstudies of the Public Order* (New York: Harper & Row, 1971), 46.
88. Eviatar Zerubavel, *Social Mindscapes*, 17.
89. See also James W. Cunningham and David W. Moore, "The Confused World of Main Idea," in James F. Baumann (ed.), *Teaching Main Idea Comprehension* (Newark, DE: International Reading Association, 1986), 10–11.
90. Elmer Luchterhand, "Knowing and Not Knowing: Involvement in Nazi Genocide," in Paul Thompson (ed.), *Our Common*

History: The Transformation of Europe (Atlantic Highlands, NJ: Humanities Press, 1982), 263; Claude Lanzmann, *Shoah: An Oral History of the Holocaust* (New York: Pantheon, 1985), 26, 97; Gordon J. Horwitz, *In the Shadow of Death: Living Outside the Gates of Mauthausen* (New York: Free Press, 1990), 27, 32, 112, 175.

91. Stanley Cohen, *States of Denial: Knowing about Atrocities and Suffering* (Cambridge: Polity, 2001), xii. See also Luchterhand, "Knowing and Not Knowing," 255; David Bankier, *The Germans and the Final Solution: Public Opinion under Nazism* (Oxford: Blackwell, 1992), 104–15, 131–32.
92. Horwitz, *In the Shadow of Death*, 35. See also 96.
93. Ibid., 94. See also Frank Graziano, *Divine Violence: Spectacle, Psychosexuality, and Radical Christianity in the Argentine "Dirty War"* (Boulder, CO: Westview, 1992), 79, 255–56; Michael Taussig, *Defacement: Public Secrecy and the Labor of the Negative* (Stanford, CA: Stanford University Press, 1999), 6, 50; Martin Amis, *Koba the Dread: Laughter and the Twenty Million* (New York: Hyperion, 2002), 39.
94. Urry, *The Tourist Gaze*, 47, 138–40; Shaul Kelner, *Tours That Bind: Diaspora, Pilgrimage, and Israeli Birthright Tourism* (New York: New York University Press, 2010), 93–94; Watson, "How Tour Operators and Travel Guidebooks Select Destinations."
95. Thomas DeGloma, personal communication.
96. Eviatar Zerubavel, *Social Mindscapes*, 32–33, 46–51. See also Park and Kitayama, "Perceiving through Culture."
97. Eviatar Zerubavel, *The Elephant in the Room: Silence and Denial in Everyday Life* (New York: Oxford University Press, 2006), 2, 20. See also Herbert Fingarette, *Self-Deception* (London: Routledge & Kegan Paul, 1969), 44; Eviatar Zerubavel, "Personal Information and Social Life" *Symbolic Interaction* 5, no. 1 (1982): 107; Shoshana Felman and Dori Laub (eds.), *Testimony: Crises of Witnessing in Literature, Psychoanalysis, and History* (New York: Routledge, 1992), 83.
98. Marjorie L. DeVault, "Producing Family Time: Practices of Leisure Activity beyond the Home," *Qualitative Sociology* 23 (2000): 492–93. Emphasis added.
99. David Grazian, "Some Animals Are More Equal Than Others: American Zoos and the Culture of Childhood" (paper presented

at the annual meeting of the American Sociological Association, New York, August 2013).
100. Ibid.
101. See also David M. Lane and Deborah A. Pearson, "The Development of Selective Attention," *Merrill-Palmer Quarterly* 28 (1982): 327–28, 333–34.
102. See, for example, Scott Paris et al., "Informed Strategies for Learning: A Program to Improve Children's Reading Awareness and Comprehension," *Journal of Educational Psychology* 76 (1984): 1242, 1251; James F. Baumann (ed.), *Teaching Main Idea Comprehension* (Newark, DE: International Reading Association, 1986); James F. Baumann, "The Direct Instruction of Main Idea Comprehension Ability," in *Teaching Main Idea Comprehension* (Newark, DE: International Reading Association, 1986), 133–78; Mark W. Aulls, "Actively Teaching Main Idea Skills," in James F. Baumann (ed.), *Teaching Main Idea Comprehension* (Newark, DE: International Reading Association, 1986), 96–132; William H. Teale and Miriam G. Martinez, "Reading Aloud to Young Children: Teachers' Reading Styles and Kindergartners' Text Comprehension," in Clotilde Pontecorvo et al. (eds.), *Children's Early Text Construction* (Mahwah, NJ: Lawrence Erlbaum, 1996), 326. See also Allan Collins and Edward E. Smith, "Teaching the Process of Reading Comprehension," in Douglas K. Detterman and Robert J. Sternberg (eds.), *How and How Much Can Intelligence Be Increased* (Norwood, NJ: Ablex, 1982), 173–85; Beth Davey, "Think Aloud: Modeling the Cognitive Processes of Reading Comprehension," *Journal of Reading* 27 (1983): 44–45; Laura R. Roehler and Gerald G. Duffy, "Direct Explanation of Comprehension Processes," in Gerald G. Duffy (ed.), *Comprehension Instruction: Perspectives and Suggestions* (New York: Longman, 1984), 265; http://www.corestandards.org/ELA-Literacy/RI/8 (accessed on August 8, 2012).
103. Pierre Bourdieu, *Distinction: A Social Critique of the Judgement of Taste* (Cambridge, MA: Harvard University Press, 1984 [1979]), 2.
104. Friedman, *Blind to Sameness*, 36–37.
105. See also Anne Fernald and Hiromi Morikawa, "Common Themes and Cultural Variations in Japanese and American

Mothers' Speech to Infants," *Child Development* 64 (1993): 651–52.
106. See also Gabriel Spitzer, "Clever Apes #29: Nature and Human Nature," http://www.wbez.org/blogs/clever-apes/2012-04/clever-apes-29-nature-and-human-nature-97867 (accessed on January 2, 2013).
107. Goodwin, "Professional Vision," 615. See also 610.
108. See, for example, Eyal M. Reingold and Neil Charness, "Perception in Chess: Evidence from Eye Movements," in Geoffrey Underwood (ed.), *Cognitive Processes in Eye Guidance* (New York: Oxford University Press, 2005), 330–33.
109. A. Mark Williams et al., "Visual Search Strategies in Experienced and Inexperienced Soccer Players," *Research Quarterly for Exercise and Sport* 65 (1994): 132; A. Mark Williams, "Perceptual Skill in Soccer: Implications for Talent Identification and Development," *Journal of Sports Sciences* 18 (2000): 742–43.
110. See, for example, Arlie Hochschild, *The Managed Heart: Commercialization of Human Feeling* (Berkeley: University of California Press, 1983), 95–114, 138–47.
111. See also Daniel Memmert, "The Effects of Eye Movements, Age, and Expertise on Inattentional Blindness," *Consciousness and Cognition* 15 (2006): 625.
112. See, for example, Calvin F. Nodine et al., "The Role of Formal Art Training on Perception and Aesthetic Judgment of Art Compositions," *Leonardo* 26 (1993): 226–27; W. H. Zangemeister et al., "Evidence for a Global Scanpath Strategy in Viewing Abstract Compared with Realistic Images," *Neuropsychologia* 33 (1995): 1009–25; Gregory Minissale, *The Psychology of Contemporary Art* (Cambridge, UK: Cambridge University Press, 2013), 102.
113. John Sloboda and Judy Edworthy, "Attending to Two Melodies at Once: The Effect of Key Relatedness," *Psychology of Music* 9 (1981): 43. See also Albert S. Bregman, *Auditory Scene Analysis: The Perceptual Organization of Sound* (Cambridge, MA: MIT Press, 1990), 42; Renaud Brochard et al., "Perceptual Organization of Complex Auditory Sequences: Effect of Number of Simultaneous Subsequences and Frequency Separation," *Journal of Experimental Psychology: Human Perception and Performance* 25 (1999): 1756.

114. See, for example, Ronald R. Mourant and Thomas H. Rockwell, "Strategies of Visual Search by Novice and Experienced Drivers," *Human Factors* 14 (1972): 331, 333–34; David Crundall et al., "Driving Experience and the Functional Field of View," *Perception* 28 (1999): 1075–87.
115. Alan Lesgold et al., "Expertise in a Complex Skill: Diagnosing X-Ray Pictures," in Michelene T. H. Chi et al. (eds.), *The Nature of Expertise* (Hillsdale, NJ: Lawrence Erlbaum Associates, 1988), 311–42; Beverly P. Wood, "Visual Expertise," *Radiology* 211 (1999): 1–3; Calvin F. Nodine and Claudia Mello-Thoms, "The Nature of Expertise in Radiology," in Jacob Beutel et al. (eds.), *Handbook of Medical Imaging, Volume 1: Physics and Psychophysics* (Bellingham, WA: SPIE Press, 2000), 859–94; Paul T. Sowden et al., "Perceptual Learning of the Detection of Features in X-Ray Images: A Functional Role for Improvements in Adults' Visual Sensitivity?" *Journal of Experimental Psychology: Human Perception and Performance* 26 (2000): 379–90.
116. Wood, "Visual Expertise," 2. See also Nodine and Mello-Thoms, "The Nature of Expertise in Radiology," 879.
117. C. Wright Mills, "Methodological Consequences of the Sociology of Knowledge," *American Journal of Sociology* 46 (1940): 322.
118. See also Kenneth V. Iserson and John C. Moskop, "Triage in Medicine, Part I: Concept, History, and Types," *Annals of Emergency Medicine* 49 (2007): 277.
119. Valerie G. A. Grossman, *Quick Reference to Triage*, 2nd ed. (Philadelphia, PA: Lippincott Williams & Wilkins, 2003), 8, 18.
120. Paul K. Saint-Amour, "Modernist Reconnaissance," *Modernism/Modernity* 10 (2003): 371.
121. See also Kurt Koffka, *Principles of Gestalt Psychology* (New York: Harcourt, Brace & World, 1935), 45; Wolfgang Metzger, *Laws of Seeing* (Cambridge, MA: MIT Press, 2006 [1936]), 5.
122. Ludwik Fleck, *Genesis and Development of A Scientific Fact* (Chicago: University of Chicago Press, 1979 [1935]); Kuhn, *The Structure of Scientific Revolutions*, 126; Arien Mack et al., "Perceptual Organization and Attention," *Cognitive Psychology* 24 (1992): 475–501.
123. Kuhn, *The Structure of Scientific Revolutions*, 111.

124. See also Georg Simmel, "The Field of Sociology," in Kurt H. Wolff (ed.), *The Sociology of Georg Simmel* (New York: Free Press, 1950 [1917]), 8.
125. See also George Lakoff and Mark Johnson, *Metaphors We Live By* (Chicago: University of Chicago Press, 1980), 163.
126. See also William Ocasio, "Towards an Attention-Based View of the Firm," *Strategic Management Journal* 18 (1997): 187–206; Thomas H. Davenport and John C. Beck, *The Attention Economy: Understanding the New Currency of Business* (Boston: Harvard Business School Press, 2001), 25–27; Bryan D. Jones and Frank R. Baumgartner, *The Politics of Attention: How Government Prioritizes Problems* (Chicago: University of Chicago Press, 2005), 232–46.
127. See, for example, Alfred Schutz, "Making Music Together: A Study in Social Relationship," *Social Research* 18 (1951): 76–97; Goffman, *Behavior in Public Places*, 24, 81–148; Adam Kendon, "The Negotiation of Context in Face-to-Face Interaction," in Alessandro Duranti and Charles Goodwin (eds.), *Rethinking Context: Language as an Interactive Phenomenon* (Cambridge: Cambridge University Press, 1992), 323–34; Heidi Keller, *Cultures of Infancy* (Mahwah, NJ: Lawrence Erlbaum Associates, 2007), 142. See also Zerubavel, *Social Mindscapes*, 97; Zerubavel, *Time Maps*, 4.
128. See also Alfred Schutz and Thomas Luckmann, *The Structures of the Life-World* (Evanston, IL: Northwestern University Press, 1973), 60, 289; Berger and Luckmann, *The Social Construction of Reality*, 45.
129. Kendon, "The Negotiation of Context in Face-to-Face Interaction," 328–29.
130. See also Roger W. Cobb and Charles D. Elder, "The Politics of Agenda-Building: An Alternative Perspective for Modern Democratic Theory," *Journal of Politics* 33 (1971): 892–915.
131. Stephen Hilgartner and Charles L. Bosk, "The Rise and Fall of Social Problems: A Public Arenas Model," *American Journal of Sociology* 94 (1988): 54.
132. Maxwell McCombs, *Setting the Agenda: The Mass Media and Public Opinion* (Cambridge, UK: Polity Press, 2004), 2.
133. Jones and Baumgartner, *The Politics of Attention*, 38.
134. Crary, *Suspensions of Perception*, 71.

135. Maxwell E. McCombs and Donald L. Shaw, "The Agenda-Setting Function of Mass Media," *Public Opinion Quarterly* 36 (1972): 176–87; McCombs, *Setting the Agenda*.
136. McCombs, *Setting the Agenda*, 1. See also Herbert J. Gans, *Deciding What's News: A Study of CBS Evening News, NBC Nightly News, Newsweek, and Time* (New York: Random House, 1979); Shanto Iyengar and Donald R. Kinder, *News That Matters: Television and American Opinion*, updated ed. (Chicago: University of Chicago Press, 2010 [1987]), 33.
137. McCombs, *Setting the Agenda*, 2. See also Iyengar and Kinder, *News That Matters*, 42–46, 113.
138. Bernard C. Cohen, *The Press and Foreign Policy* (Princeton, NJ: Princeton University Press, 1963), 13. Emphasis added. See also Timur Kuran, *Private Truths, Public Lies: The Social Consequences of Preference Falsification* (Cambridge, MA: Harvard University Press, 1995), 187.
139. Anthony Downs, "Up and Down with Ecology: The 'Issue-Attention Cycle,'" *The Public Interest* 28 (1972): 38. See also Iyengar and Kinder, *News That Matters*, 33; Eric Klinenberg, *Heat Wave: A Social Autopsy of Disaster in Chicago* (Chicago: University of Chicago Press, 2002), 221–24.
140. John W. Kingdon, *Agendas, Alternatives, and Public Policies* (New York: HarperCollins, 1984), 99–105.
141. See, for example, Juleyka Lantigua-Williams, "Missing & Black: Helping Families Cope," http://jetmag.com/news/missing-black-helping-families-cope (accessed on May 29, 2013).
142. See, for example, Joshua Meyrowitz, "The Press Rejects a Candidate," *Columbia Journalism Review*, March/April 1992, 46–47.

Chapter 5

1. Marion Milner, *A Life of One's Own* (London: Routledge, 2011 [1934]), 111.
2. Alexandra Horowitz, *On Looking: Eleven Walks with Expert Eyes* (New York: Scribner, 2013), 26.
3. Aldous Huxley, "The Doors of Perception," in *The Doors of Perception and Heaven and Hell* (New York: HarperCollins, 2009 [1954]), 23.

4. D. E. Berlyne, *Aesthetics and Psychobiology* (New York: Appleton-Century-Crofts, 1971), 100.
5. Andrew McGhie and James Chapman, "Disorders of Attention and Perception in Early Schizophrenia," *British Journal of Medical Psychology* 34 (1961): 110.
6. See also Horowitz, *On Looking*, 11.
7. Charlie D. Broad, in Huxley, "The Doors of Perception," 22–23.
8. Daniel Memmert and Philip Furley, "'I Spy with My Little Eye!': Breadth of Attention, Inattentional Blindness, and Tactical Decision Making in Team Sports," *Journal of Sport and Exercise Psychology* 29 (2007): 376. Emphasis added.
9. Eugen Bleuler, *Dementia Praecox or the Group of Schizophrenias* (Madison, CT: International Universities Press, 1950 [1911]), 68; McGhie and Chapman, "Disorders of Attention and Perception in Early Schizophrenia," 112–13; Thomas Freeman et al., *Studies on Psychosis: Descriptive, Psycho-analytic, and Psychological Aspects* (New York: International Universities Press, 1966), 73, 78, 190–91; Olga Bogdashina, *Sensory Perceptual Issues in Autism and Asperger Syndrome: Different Sensory Experiences—Different Perceptual Worlds* (London: Jessica Kingsley Publishers, 2003), 48.
10. Bleuler, *Dementia Praecox or the Group of Schizophrenias*, 68.
11. McGhie and Chapman, "Disorders of Attention and Perception in Early Schizophrenia," 112.
12. Andrew McGhie, *Pathology of Attention* (Baltimore, MD: Penguin Books, 1969), 51–54.
13. Norma MacDonald, "Living with Schizophrenia," *Canadian Medical Association Journal* 82 (1960): 218–19. Emphasis added. See also McGhie and Chapman, "Disorders of Attention and Perception in Early Schizophrenia," 104–05.
14. McGhie and Chapman, "Disorders of Attention and Perception in Early Schizophrenia," 113, 105.
15. Ibid., 113.
16. Marcel Kinsbourne and Paula J. Caplan, *Children's Learning and Attention Problems* (Boston: Little, Brown, and Co., 1979), 3. Emphasis added.
17. See also Herbert J. Schlesinger, "Cognitive Attitudes in Relation to Susceptibility to Interference," *Journal of Personality* 22 (1954): 356.

18. Aidan Moran, "Attention in Sport," in Stephen D. Mellalieu and Sheldon Hanton (eds.), *Advances in Applied Sport Psychology: A Review* (London: Routledge, 2009), 197.
19. See, for example, Memmert and Furley, "'I Spy with My Little Eye!'" 2007, 366.
20. See also Edwin S. Shneidman, *The Suicidal Mind* (New York: Oxford University Press, 1996), 59–60.
21. William Blake, *The Marriage of Heaven and Hell* (London: Oxford University Press, 1975 [1794]), Plate 14 and p. xxii.
22. Milner, *A Life of One's Own*, 111.
23. Ruth Silver, *Invisible: My Journey through Vision and Hearing Loss* (iUniverse, 2012), 3. See also Leonard Shlain, *The Alphabet versus the Goddess: The Conflict between Word and Image* (New York: Penguin Arkana, 1998), 24, 26, 43.
24. See also Elliot G. Mishler, "Meaning in Context: Is There Any Other Kind?" *Harvard Educational Review* 49 (1979): 2, 17; Dale Spender, *Man Made Language* (London: Routledge & Kegan Paul, 1980), 165; Janice Haaken, "Field Dependence Research: A Historical Analysis of a Psychological Construct," *Signs* 13 (1988): 318; Eviatar Zerubavel, *The Fine Line: Making Distinctions in Everyday Life* (Chicago: University of Chicago Press, 1993 [1991]), 116; Martin J. Doherty et al., "The Context-Sensitivity of Visual Size Perception Varies across Cultures," *Perception* 37 (2008): 1426–33.
25. See also Zerubavel, *The Fine Line*, 116.
26. See, for example, Eviatar Zerubavel, *The Elephant in the Room: Silence and Denial in Everyday Life* (New York: Oxford University Press, 2006), 82; Les Fehmi and Jim Robbins, *The Open-Focus Brain: Harnessing the Power of Attention to Heal Mind and Body* (Boston: Trumpeter, 2008), 136.
27. See, for example, Eviatar Zerubavel, *Social Mindscapes: An Invitation to Cognitive Sociology* (Cambridge, MA: Harvard University Press, 1997), 52; Fehmi and Robbins, *The Open-Focus Brain*, 23; Steven C. Hayes et al., "Experiential Avoidance and Behavioral Disorders: A Functional Dimensional Approach to Diagnosis and Treatment," *Journal of Consulting and Clinical Psychology* 64 (1996): 1157–58.
28. See also Patrick Cavanagh and George A. Alvarez, "Tracking Multiple Targets with Multifocal Attention," *Trends in Cognitive Sciences* 9 (2005): 349–54.

29. See, for example, Anne M. Treisman, "How the Deployment of Attention Determines What We See," *Visual Cognition* 14 (2006): 411–43; Narayanan Srinivasan et al., "Focused and Distributed Attention," *Progress in Brain Research* 176 (2009): 87–100; Sang C. Chong and Karla K. Evans, "Distributed vs. Focused Attention (Count vs. Estimate)," *Wiley Interdisciplinary Reviews: Cognitive Science* 2 (2011): 634–38.
30. See also Anne M. Treisman, "Strategies and Models of Selective Attention," *Psychological Review* 76 (1969): 282–99; Zenon W. Pylyshyn and Ron W. Storm, "Tracking Multiple Independent Targets: Evidence for a Parallel Tracking Mechanism," *Spatial Vision* 3 (1988): 179–97.
31. See, for example, Dario D. Salvucci and Niels A. Taatgen, *The Multitasking Mind* (New York: Oxford University Press, 2011).
32. See, for example, Leo Gugerty, "Situation Awareness in Driving," in Donald L. Fisher et al. (eds.), *Handbook of Driving Simulation for Engineering, Medicine, and Psychology* (Boca Raton, FL: CRC Press, 2011), 19.265–19.272.
33. See Zerubavel, *The Fine Line*, 34.
34. On the distinction between those two modes of processing information, see, for example, James T. Townsend, "Serial vs. Parallel Processing: Sometimes They Look like Tweedledum and Tweedledee But They Can (and Should) Be Distinguished," *Psychological Science* 1 (1990): 46–54.
35. Paul K. Saint-Amour, "Modernist Reconnaissance," *Modernism/Modernity* 10 (2003): 352.
36. Sigfried Giedion, *Space, Time, and Architecture: The Growth of A New Tradition*, 3rd ed. (Cambridge, MA: Harvard University Press, 1956), 433–34; Anton Ehrenzweig, *The Hidden Order of Art: A Study in the Psychology of Artistic Imagination* (Berkeley: University of California Press, 1971 [1967]), 67; Stephen Kern, *The Culture of Time and Space 1880–1918* (Cambridge, MA: Harvard University Press, 1983), 195–96.
37. See also Anton Ehrenzweig, *The Psycho-Analysis of Artistic Vision and Hearing: An Introduction to a Theory of Unconscious Perception* (New York: The Julian Press, 1953), 41; Ehrenzweig, *The Hidden Order of Art*, 25; Albert S. Bregman, *Auditory Scene Analysis: The Perceptual Organization of Sound* (Cambridge, MA: MIT Press, 1990), 494; E. Bigand et al., "Divided

Attention in Music," *International Journal of Psychology* 35 (2000): 271.
38. Aaron Copland, *What to Listen for in Music*, rev. ed. (New York: McGraw-Hill, 1957 [1939]), 106. See also John Sloboda and Judy Edworthy, "Attending to Two Melodies at Once: The Effect of Key Relatedness," *Psychology of Music* 9 (1981): 39.
39. Ehrenzweig, *The Psycho-Analysis of Artistic Vision and Hearing*, 42. Emphasis added.
40. See also Joseph Jastrow, *The Subconscious* (Boston: Houghton, Mifflin & Co., 1906), 53.
41. Harris M. Berger, *Metal, Rock, and Jazz: Perception and the Phenomenology of Musical Experience* (Middletown, CT: Wesleyan University Press, 1999), 124. Emphasis added.
42. See also Aron Gurwitsch, *The Field of Consciousness* (Pittsburgh, PA: Duquesne University Press, 1964); P. Sven Arvidson, *The Sphere of Attention: Context and Margin* (Dordrecht, The Netherlands: Springer, 2006).
43. Francisco J. Varela and Natalie Depraz, "Wisdom Traditions and the Ways of Reduction," in Natalie Depraz et al. (eds.), *On Becoming Aware: A Pragmatics of Experiencing* (Amsterdam, The Netherlands: John Benjamins Publishing Company, 2003), 217. See also Drew Leder, *The Absent Body* (Chicago: University of Chicago Press, 1990), 11.
44. See, for example, Bela Julesz, "Perceptual Limits of Texture Discrimination and Their Implications to Figure-Ground Separation," in Emanuel L. J. Leeuwenberg and H. F. J. M. Buffart (eds.), *Formal Theories of Visual Perception* (New York: John Wiley & Sons, 1978), 211. See also Syoichi Iwasaki, "Spatial Attention and Two Modes of Visual Consciousness," *Cognition* 49 (1993): 211–33.
45. See, for example, Kurt Koffka, *Principles of Gestalt Psychology* (New York: Harcourt, Brace & World, 1935), 201–02, 204; Shlain, *The Alphabet versus the Goddess*, 24–25.
46. See, for example, Varela and Depraz, "Wisdom Traditions and the Ways of Reduction," 217. See also Eugene Halton, "Peircean Animism and the End of Civilization," *Contemporary Pragmatism* 2 (2005): 156.
47. See, for example, Tom Brown, *Tom Brown's Field Guide to Nature Observation and Tracking* (New York: Berkley Books, 1983),

39–41, 62; Igor Kusakov, "Deconcentration of Attention: Addressing the Complexity of Software Engineering," January 27, 2012, http://deconcentration-of-attention.com/ (accessed on July 13, 2012). See also George S. Day and Paul J. H. Schoemaker, *Peripheral Vision: Detecting the Weak Signals That Will Make or Break Your Company* (Boston: Harvard Business School Press, 2006).

48. See, for example, David Crundall et al., "Driving Experience and the Functional Field of View," *Perception* 28 (1999): 1075–87; http://www.ehow.com/video_4414788-use-peripheral-vision-juggling.html (accessed on May 31, 2013); Vincent Nougier et al., "Information Processing in Sport and 'Orienting of Attention,'" *International Journal of Sport Psychology* 22 (1991): 313; Caterina P. Pesce-Anzeneder and Rainer Bösel, "Modulation of the Spatial Extent of the Attentional Focus in High-Level Volleyball Players," *European Journal of Cognitive Psychology* 10 (1998): 247–67; Memmert and Furley, "'I Spy with My Little Eye!'"; Dave Jones, *Basketball—It's All about the Shot*, 73–79, http://www.basketballshootingcoach.com/files/1625141/uploaded/BBShootingBookPDF2.pdf (accessed on May 30, 2013).

49. Memmert and Furley, "'I Spy with My Little Eye!'" 367.

50. James H. Austin, *Mediating Selflessly: Practical Neural Zen* (Cambridge, MA: MIT Press, 2011), 19–20.

51. Thom Hartmann, *Attention Deficit Disorder: A Different Perception* (Grass Valley, CA: Underwood Books, 1997 [1993]), 151–52. Emphasis added.

52. See also Jeffrey B. Rubin, "Deepening Listening: The Marriage of Buddha and Freud," in Uwe P. Gielen et al. (eds.), *Principles of Multicultural Counseling and Therapy* (New York: Routledge, 2008), 378; James H. Austin, *Zen-Brain Reflections: Reviewing Recent Developments in Meditation and States of Consciousness* (Cambridge, MA: MIT Press, 2006), 30.

53. Varela and Depraz, "Wisdom Traditions and the Ways of Reduction," 221.

54. Livia Kohn, "Meditation and Visualization," in Fabrizio Pregadio (ed.), *The Encyclopedia of Taoism* (New York: Routledge, 2008), vol. 1, 118. Emphasis added. See also Austin, *Zen-Brain Reflections*, 30; James H. Austin, *Selfless Insight: Zen and the Meditative Transformations of Consciousness* (Cambridge, MA: MIT Press, 2009), 4.

55. Varela and Depraz, "Wisdom Traditions and the Ways of Reduction," 222.
56. Bruce R. Dunn et al. "Concentration and Mindfulness Meditations: Unique Forms of Consciousness?" *Applied Psychophysiology and Biofeedback* 24 (1999): 148. Emphasis added. See also 164.
57. See, for example, Jon Kabat-Zinn, *Full Catastrophe Living: Using the Wisdom of Your Body and Mind to Face Stress, Pain, and Illness* (New York: Delacorte Press, 1990), 2, 20, 24–25, 65–66; Victoria M. Follette et al., "Acceptance, Mindfulness, and Trauma," in Steven C. Hayes et al. (eds.), *Mindfulness and Acceptance: Expanding the Cognitive-Behavioral Tradition* (New York: Guilford, 2004), 199; Austin, *Zen-Brain Reflections*, 30.
58. Ronald S. Friedman et al., "Attentional Priming Effects on Creativity," *Creativity Research Journal* 15 (2003): 277–86.
59. See also Margaret Dykes and Andrew McGhie, "A Comparative Study of Attentional Strategies of Schizophrenic and Highly Creative Normal Subjects," *British Journal of Psychiatry* 128 (1976): 52, 55; Memmert and Furley, "'I Spy with My Little Eye!'"
60. See, for example, Joseph Kasof, "Creativity and Breadth of Attention," *Creativity Research Journal* 10 (1997): 303–04.
61. Friedman et al., "Attentional Priming Effects on Creativity," 278. See also 279.
62. Shelley H. Carson et al., "Decreased Latent Inhibition Is Associated with Increased Creative Achievement in High-Functioning Individuals," *Journal of Personality and Social Psychology* 85 (2003): 505.
63. See also ibid., 499.
64. Marie Dellas and Eugene L. Gaier, "Identification of Creativity: The Individual," *Psychological Bulletin* 73 (1970): 62. Emphasis added. See also 61.
65. Dykes and McGhie, "A Comparative Study of Attentional Strategies of Schizophrenic and Highly Creative Normal Subjects," 55. Emphasis added.
66. See also Roy Schafer and Gardner Murphy, "The Role of Autism in a Visual Figure-Ground Relationship," *Journal of Experimental Psychiatry* 32 (1943): 336.

67. See also James J. Gibson, *The Perception of the Visual World* (Westport, CT: Greenwood Press, 1950), 39.
68. Wolfgang Köhler, *Gestalt Psychology: An Introduction to New Concepts in Modern Psychology* (New York: Mentor Books, 1947), 106–07.
69. Koffka, *Principles of Gestalt Psychology*, 194–95; Rudolf Arnheim, "The Perception of Maps," in *New Essays on the Psychology of Art* (Berkeley: University of California Press, 1986 [1976]), 199. See also Gaetano Kanizsa, "Perception, Past Experience, and the 'Impossible Experiment,'" in *Organization in Vision: Essays on Gestalt Perception* (New York: Praeger, 1979 [1969]), 45–46.
70. See, for example, Schlesinger, "Cognitive Attitudes in Relation to Susceptibility to Interference," 356; MacDonald, "Living with Schizophrenia," 219; McGhie and Chapman, "Disorders of Attention and Perception in Early Schizophrenia," 113; Dykes and McGhie, "A Comparative Study of Attentional Strategies of Schizophrenic and Highly Creative Normal Subjects," 51; David M. Lane and Deborah A. Pearson, "The Development of Selective Attention," *Merrill-Palmer Quarterly* 28 (1982): 333–34; Ronald A. Rensink et al., "To See or Not to See: The Need for Attention to Perceive Changes in Scenes," *Psychological Science* 8 (1997): 372; Kasof, "Creativity and Breadth of Attention," 303; Charles Kostelnick and David D. Roberts. *Designing Visual Language: Strategies for Professional Communicators* (Boston: Allyn and Bacon, 1998), 50; Steven S. Hsiao and Francisco Vega-Bermudez, "Attention in the Somatosensory System," in Randall J. Nelson (ed.), *The Somatosensory System: Deciphering the Brain's Own Body Image* (Boca Raton, FL: CRC Press LLC, 2001), 198.
71. See, for example, Scott Paris et al., "Informed Strategies for Learning: A Program to Improve Children's Reading Awareness and Comprehension," *Journal of Educational Psychology* 76 (1984): 1242, 1251; James F. Baumann (ed.), *Teaching Main Idea Comprehension* (Newark, DE: International Reading Association, 1986); James F. Baumann, "The Direct Instruction of Main Idea Comprehension Ability," in *Teaching Main Idea Comprehension*, 133–78; Mark W. Aulls, "Actively Teaching Main Idea Skills," in *Teaching Main Idea Comprehension*, 96–132; James W. Cunningham and David W. Moore, "The Confused World

of Main Idea," in *Teaching Main Idea Comprehension*, 10–11; William H. Teale and Miriam G. Martinez, "Reading Aloud to Young Children: Teachers' Reading Styles and Kindergartners' Text Comprehension," in Clotilde Pontecorvo et al. (eds.), *Children's Early Text Construction* (Mahwah, NJ: Lawrence Erlbaum, 1996), 326, 336–38.

72. Suzanne Fleischman, "Discourse Functions of Tense-Aspect Oppositions in Narrative: Toward a Theory of Grounding," *Linguistics* 23 (1985): 858.
73. Kinsbourne and Caplan, *Children's Learning and Attention Problems*, 3–4, 10.
74. Notice in this regard the subtitle of Thomas E. Brown's book *Attention Deficit Disorder: The Unfocused Mind in Children and Adults* (New Haven, CT: Yale University Press, 2005).
75. See also Jonathan Crary, *Suspensions of Perception: Attention, Spectacle, and Modern Culture* (Cambridge, MA: MIT Press, 1999), 35; Rosemarie Garland-Thomson, *Staring: How We Look* (New York: Oxford University Press, 2009), 22.
76. See also Julian Silverman, "Scanning-Control Mechanism and 'Cognitive Filtering' in Paranoid and Non-Paranoid Schizophrenia," *Journal of Consulting Psychology* 28 (1964): 391.
77. See, for example, Ami Klin et al., "Visual Fixation Patterns During Viewing of Naturalistic Social Situations as Predictors of Social Competence in Individuals with Autism," *Archives of General Psychiatry* 59 (2002): 809–16; Warren Jones and Ami Klin, "Attention to Eyes Is Present but in Decline in 2-6-Month-Old Infants Later Diagnosed with Autism," *Nature* 12715 (November 6, 2013). See also Bogdashina, *Sensory Perceptual Issues in Autism and Asperger Syndrome*, 102.
78. Bogdashina, *Sensory Perceptual Issues in Autism and Asperger Syndrome*, 101.
79. See, for example, Francesca Happé, "Autism: Cognitive Deficit or Cognitive Style?" *Trends in Cognitive Sciences* 3 (1999): 217; Pier Jaarsma and Stellan Welin, "Autism as a Natural Human Variation: Reflections on the Claims of the Neurodiversity Movement," *Health Care Analysis* February 11, 2011, http://www.imh.liu.se/avd_halsa_samhalle/filarkiv1/1.264263/JaarsmaWelin2011Autismasanaturalvariation.pdf (accessed on April 7, 2014).

80. Therese Jolliffe and Simon Baron-Cohen, "Are People with Autism and Asperger Syndrome Faster Than Normal on the Embedded Figures Test?" *Journal of Child Psychology and Psychiatry* 38 (1997): 527.
81. See, for example, Amitta Shah and Uta Frith, "An Islet of Ability in Autistic Children: A Research Note," *Journal of Child Psychology and Psychiatry and Allied Disciplines* 24 (1983): 613–20; Leon K. Miller, *Musical Savants: Exceptional Skill in the Mentally Retarded* (Hillsdale, NJ: Lawrence Erlbaum, 1989), 51–56; Jeanne Townsend and Eric Courchesne, "Parietal Damage and Narrow 'Spotlight' Spatial Attention," *Journal of Cognitive Neuroscience* 6 (1994): 220; Uta Frith and Francesca Happé, "Autism: Beyond 'Theory of Mind,'" *Cognition* 50 (1994): 121–22; Francesca Happé, *Autism: An Introduction to Psychological Theory* (Cambridge, MA: Harvard University Press, 1995), 119; Jolliffe and Baron-Cohen, "Are People with Autism and Asperger Syndrome Faster Than Normal on the Embedded Figures Test?"; Bogdashina, *Sensory Perceptual Issues in Autism and Asperger Syndrome*, 65–67, 102; Pamela F. Heaton, "Pitch Memory, Labelling, and Disembedding in Autism," *Journal of Child Psychology and Psychiatry* 44 (2003): 543–51; Temple Grandin and Catherine Johnson, *Animals in Translation: Using the Mysteries of Autism to Decode Animal Behavior* (Orlando, FL: Harvest Books, 2006 [2005]), 296–97; Temple Grandin and Richard Panek, *The Autistic Brain: Thinking Across the Spectrum* (Boston: Houghton Mifflin Harcourt, 2013), 122.
82. Grandin and Johnson, *Animals in Translation*, 296–97.
83. Ibid., 297–98.
84. See also Eviatar Zerubavel, *The Elephant in the Room: Silence and Denial in Everyday Life* (New York: Oxford University Press, 2006), 65–68.
85. Thomas S. Kuhn, *The Structure of Scientific Revolutions*, 2nd ed. (Chicago: University of Chicago Press, 1970 [1962]), 85, 111–14, 150.
86. See also Oleg Bakhtiarov, "Psychonetics," August 13, 2009, http://www.psychotechnology.ru/publication/item56.html (accessed on July 13, 2012); Kusakov, "Deconcentration of Attention."

87. Peggy DesAutels, "Gestalt Shifts in Moral Perception," in Larry May et al. (eds.), *Mind and Morals: Essays on Cognitive Science and Ethics* (Cambridge, MA: MIT Press, 1996), 134. Emphasis added.
88. John Lennon, "Beautiful Boy."
89. See, for example, Zerubavel, *The Elephant in the Room*, 61–72.
90. See, for example, Paul Ekman and Wallace V. Friesen, "Nonverbal Leakage and Clues to Deception," *Psychiatry* 32 (1969): 88–106.
91. See http://www.huffingtonpost.co.uk/2012/01/23/slovakian-violinist-lukas-kmit-nokia-ringtone_n_1223086.html (accessed on January 11, 2014).
92. Mark Olsen, "'20 Feet from Stardom' Moves Spotlight to the Background," *Los Angeles Times*, June 18, 2013, http://articles.latimes.com/2013/jun/18/entertainment/la-et-mn-20-feet-from-stardom-20130619 (accessed on February 2, 2014).
93. Arthur J. Deikman, "De-Automatization and the Mystic Experience," *Psychiatry* 29 (1966): 324–38. See also Victor Shklovsky, "Art as Technique," in Lee T. Lemon and Marion J. Reis (eds.), *Russian Formalist Criticism: Four Essays* (Lincoln: University of Nebraska Press, 1965 [1917]), 5–24; Robert Ornstein, *Meditation and Modern Psychology* (Los Altos, CA: Malor Books, 2008 [1971]), 61–76; Willie Van Peer, *Stylistics and Psychology: Investigations of Foregrounding* (London: Croom Helm, 1986), 2; Kabat-Zinn, *Full Catastrophe Living*, 21–22.
94. Tom Brown, "Fill Your Senses, Light Up Your Life," *Reader's Digest*, August 1984, 154–56.
95. Seth S. Horowitz, "The Science and Art of Listening," *New York Times*, November 11, 2012, http://www.nytimes.com/2012/11/11/opinion/sunday/why-listening-is-so-much-more-than-hearing.html?_r=0 (accessed April 22, 2013).
96. Horowitz, *On Looking*, 15.
97. Shklovsky, "Art as Technique"; Ornstein, *Meditation and Modern Psychology*, 61–76; Van Peer, *Stylistics and Psychology*, 1–5; David S. Miall and Don Kuiken, "Foregrounding, Defamiliarization, and Affect: Response to Literary Stories," *Poetics* 22 (1994): 391.
98. See also Wayne H. Brekhus, "A Sociology of the Unmarked: Redirecting Our Focus," *Sociological Theory* 16 (1998): 34–51.

99. Horace Miner, "Body Ritual among the Nacirema," *American Anthropologist* 58 (1956): 503–07.
100. See, for example, Erving Goffman, *Behavior in Public Places: Notes on the Social Organization of Gatherings* (New York: Free Press, 1963); Harold Garfinkel, "Studies of the Routine Grounds of Everyday Activities," in *Studies in Ethnomethodology* (Englewood Cliffs, NJ: Prentice-Hall, 1967 [1964]), 35–75. See also Brekhus, "A Sociology of the Unmarked," 44.
101. Thomas Schelling, in a publicity blurb for Goffman's book *Strategic Interaction* that appears on the back cover of the paperback edition of William Labov's book *Sociolinguistic Patterns*.
102. See also Herbert Fingarette, *Self-Deception* (London: Routledge & Kegan Paul, 1969), 39–51; Larraine Segil, *Dynamic Leader, Adaptive Organization: Ten Essential Traits for Managers* (New York: John Wiley & Sons, 2002), 125.
103. Genesis 3:5–7.
104. See also Brekhus, "A Sociology of the Unmarked," 45; Peter Stockwell, *Cognitive Poetics: An Introduction* (London: Routledge, 2002), 14.
105. Arthur C. Doyle, "Silver Blaze," in *Sherlock Holmes: The Complete Novels and Stories*, vol. 1 (New York: Bantam Books, 1986 [1892]), 540.
106. Edmund Husserl, *Thing and Space: Lectures of 1907* (Dordrecht, The Netherlands: Kluwer Academic Publishers, 2010 [1907]), 323.
107. Brown, *Tom Brown's Field Guide to Nature Observation and Tracking*, 47. Emphasis added.
108. Fehmi and Robbins, *The Open-Focus Brain*, 157–59. Emphasis added.
109. Ibid., 69.
110. Richard D. Zakia, *Perception and Imaging*, 2nd ed. (Boston: Focal Press, 2002), 21. Emphasis added.
111. Betty Edwards, *Drawing on the Right Side of the Brain: A Course in Enhancing Creativity and Artistic Confidence* (Los Angeles: J. P. Tarcher, 1979), 102–09; Betty Edwards, *The New Drawing on the Right Side of the Brain* (New York: Jeremy P. Tarcher/Putnam, 1999), 116–35.
112. Edwards, *The New Drawing on the Right Side of the Brain*, 118. See also Ehrenzweig, *The Psycho-Analysis of Artistic Vision and Hearing*, 28, 36.

113. Carl Purcell, *Your Artist's Brain* (Cincinnati, OH: North Light Books, 2010), 141–42.
114. See also Jon Driver and Gordon C. Baylis, "Edge-Assignment and Figure-Ground Segmentation in Short-Term Visual Matching," *Cognitive Psychology* 31 (1996): 249–50; Zakia, *Perception and Imaging*, 13.
115. Edwards, *The New Drawing on the Right Side of the Brain*, 95.
116. Ibid., 119.
117. See, for example, Figure 15b in Wolfgang Metzger, *Laws of Seeing* (Cambridge, MA: MIT Press, 2006 [1936]), 12.
118. See also Ehrenzweig, *The Hidden Order of Art*, 22.
119. Roger N. Shepard, *Mind Sights: Original Visual Illusions, Ambiguities, and Other Anomalies, with a Commentary on the Play of Mind in Perception and Art* (New York: W. H. Freeman and Company, 1990), 72.
120. Gyorgy Kepes, *Language of Vision* (Chicago: Paul Theobald, 1951), 102.
121. Doris Schattschneider, *Visions of Symmetry: Notebooks, Periodic Drawings, and Related Work of M. C. Escher* (New York: W. H. Freeman, 1990), 296.
122. Maurits C. Escher, *Escher on Escher: Exploring the Infinite* (New York: Harry N. Abrams, 1989 [1986]), 35. See also 83–84, 93–94; Douglas R. Hofstadter, *Gödel, Escher, Bach: An Eternal Golden Braid* (New York: Basic Books, 1979), 67.
123. See also Schattschneider, *Visions of Symmetry*, 262, 264.
124. See also David Chevan, "The Double Bass as a Solo Instrument in Early Jazz," *The Black Perspective in Music* 17 (1989): 73–91.
125. Tom Stoppard, *Rosencrantz and Guildenstern Are Dead* (New York: Samuel French, 1995 [1967]).
126. See also Kern, *The Culture of Time and Space 1880–1918*, 179.
127. Ibid., 155.
128. See, for example, Roger Trancik, *Finding Lost Space: Theories of Urban Design* (New York: John Wiley & Sons, 1986), 100, 106.
129. See, for example, Stéphane Mallarmé, *A Tomb for Anatole* (San Francisco: North Point, 1983 [1961]), 106, 131. See also Kern, *The Culture of Time and Space 1880–1918*, 172–74.
130. See, for example, Jackson Mac Low, *Representative Works 1938—1985* (New York: Roof, 1986), 53, 62–70, 106–27.
131. Wolfgang Metzger, *Laws of Seeing*, 4–5.

132. Koffka, *Principles of Gestalt Psychology*, 208.
133. See also ibid., 184; Gibson, *The Perception of the Visual World*, 38; Fehmi and Robbins, *The Open-Focus Brain*, 158; Purcell, *Your Artist's Brain*, 141.
134. See also Wafu Teshigahara, *Ikebana: The New Illustrated Guide to Mastery* (Tokyo: Kodansha International, 1966); Zakia, *Perception and Imaging*, 12.
135. See, for example, Dorr Bothwell and Marlys Mayfield, *Notan: The Dark-Light Principle of Design* (New York: Dover Publications, 1991 [1968]), 8, 16. See also Deganit S. Schocken, *How Many Is One* (Tel Aviv, Israel: Tel Aviv Museum of Art, 2003).
136. Hugh B. Cott, *Adaptive Coloration in Animals* (London: Methuen, 1957 [1940]), 4; Ehrenzweig, *The Hidden Order of Art*, 11; Purcell, *Your Artist's Brain*, 93.
137. Purcell, *Your Artist's Brain*, 85.
138. Ehrenzweig, *The Psycho-Analysis of Artistic Vision and Hearing*, 38.
139. See also Harry Helson, "The Psychology of Gestalt," *American Journal of Psychology* 36 (1925): 496; Gibson, *The Perception of the Visual World*, 39; Kepes, *Language of Vision*, 60; Heinz Werner and Seymour Wapner, "Toward a General Theory of Perception," in David C. Beardslee and Michael Wertheimer (eds.), *Readings in Perception* (Princeton, NJ: D. Van Nostrand Company, 1958 [1952]), 507; Ehrenzweig, *The Psycho-Analysis of Artistic Vision and Hearing*, viii; Ehrenzweig, *The Hidden Order of Art*, 8, 22; Fleischman, "Discourse Functions of Tense-Aspect Oppositions in Narrative," 854.
140. See Rudolf Arnheim, *The Power of the Center: A Study of Composition in the Visual Arts* (Berkeley: University of California Press, 1982).
141. See also Rudolf Arnheim, *Art and Visual Perception: A Psychology of the Creative Eye* (Berkeley: University of California Press, 1967 [1954]), 231; Teddy Brunius, "Inside and Outside the Frame of a Work of Art," in *Idea and Form: Studies in the History of Art* [Stockholm: Almqvist and Wiksell (Uppsala Studies in the History of Art, Figura Nova Series, no.1), 1959], 15–16; Arnheim, *The Power of the Center*, 58–59.
142. See also Michel Butor, "Mondrian: The Square and Its Inhabitant," in *Inventory: Essays* (New York: Simon and Schuster, 1968 [1965]), 239; Arnheim, *The Power of the Center*, 150–51.

143. See, for example, Edward E. Cummings, *A Selection of Poems* (New York: Harvest, 1965), 140. See also Barbara H. Smith, *Poetic Closure: A Study of How Poems End* (Chicago: University of Chicago Press, 1968), 237, 246–50.
144. Luigi Pirandello, *Tonight We Improvise* (New York: Samuel French, 1960 [1932]); Clifford Odets, "Waiting for Lefty," in *Waiting for Lefty and Other Plays* (New York: Grove Press, 1993 [1935]), 5–31; Joseph Heller, *We Bombed in New Haven* (New York: Delta, 1967). See also Richard Kostelanetz, *The Theater of Mixed Means* (New York: Dial Press, 1968), 23–25; Richard Schechner, *Environmental Theater* (New York: Hawthorn, 1973), 4, 7, 35–39, 49–53, 59; Boris Uspensky, *A Poetics of Composition* (Berkeley: University of California Press, 1973), 139; Yi-Fu Tuan, *Segmented Worlds and Self* (Minneapolis: University of Minnesota Press, 1982), 189–92; Kern, *The Culture of Time and Space 1880–1918*, 199–201; Michael Kirby and Victoria N. Kirby, *Futurist Performance* (New York: PAJ Publications, 1986), 48–69.
145. See also Michael Kirby (ed.). *Happenings: An Illustrated Anthology* (New York: E. P. Dutton, 1965), 24; Allan Kaprow, *Assemblage, Environments, and Happenings* (New York: Harry N. Abrams, 1966), 154, 160, 165.
146. See also Ehrenzweig, *The Psycho-Analysis of Artistic Vision and Hearing*, 30; Ehrenzweig, *The Hidden Order of Art*, 119; Brown, *Tom Brown's Field Guide to Nature Observation and Tracking*, 62; Austin, *Zen-Brain Reflections*, 30.
147. See also Marion Milner, "The Framed Gap," in *The Suppressed Madness of Sane Men: Forty-Four Years of Exploring Psychoanalysis* (London: Tavistock, 1987 [1952]), 81; Ehrenzweig, *The Psycho-Analysis of Artistic Vision and Hearing*, 53; Ehrenzweig, *The Hidden Order of Art*, 21, 28–30, 107.
148. See also Ehrenzweig, *The Psycho-Analysis of Artistic Vision and Hearing*, vii-viii, 4, 30, 33, 42; Ehrenzweig, *The Hidden Order of Art*, 23; Charles Fisher, "Further Observations on the Poetzl Phenomenon: The Effects of Subliminal Visual Stimulation on Dreams, Images, and Hallucinations," *Psychoanalysis and Contemporary Thought* 11 (1988): 38–39.
149. See also Arthur J. Deikman, "Bimodal Consciousness," *Archives of General Psychiatry* 25 (1971): 481–82.

150. Eviatar Zerubavel, "Horizons: On the Sociomental Foundations of Relevance," *Social Research* 60 (1993): 397–413.
151. For how the world looks to people with actual loss of peripheral vision, see, for example, http://www.sciencephoto.com/media/101371/enlarge (accessed on May 11, 2012).
152. See, for example, Milner, *A Life of One's Own*; Day and Schoemaker, *Peripheral Vision*; Fehmi and Robbins, *The Open-Focus Brain*.
153. William James, *The Principles of Psychology* (Cambridge, MA: Harvard University Press, 1983 [1890]), 401.

BIBLIOGRAPHY

Ackerman, Diane. *A Natural History of the Senses.* New York: Random House, 1990.

Alves, Stephanie E. "Blind Trusting the Blind: The Attentional Norms of Social Trust." Unpublished manuscript, Rutgers University, New Brunswick, NJ, 2013.

Amis, Martin. *Koba the Dread: Laughter and the Twenty Million.* New York: Hyperion, 2002.

Arnheim, Rudolf. *Art and Visual Perception: A Psychology of the Creative Eye.* Berkeley: University of California Press, 1967 [1954].

———. "The Perception of Maps." In *New Essays on the Psychology of Art,* 194–202. Berkeley: University of California Press, 1986 [1976].

———. *The Power of the Center: A Study of Composition in the Visual Arts.* Berkeley: University of California Press, 1982.

———. *Visual Thinking.* Berkeley: University of California Press, 1969.

Arvidson, P. Sven. *The Sphere of Attention: Context and Margin.* Dordrecht, The Netherlands: Springer, 2006.

Aulls, Mark W. "Actively Teaching Main Idea Skills." In James F. Baumann (ed.), *Teaching Main Idea Comprehension,* 96–132. Newark, DE: International Reading Association, 1986.

Austin, James H. *Meditating Selflessly: Practical Neural Zen.* Cambridge, MA: MIT Press, 2011.

Austin, James H. *Selfless Insight: Zen and the Meditative Transformations of Consciousness.* Cambridge, MA: MIT Press, 2009.

———. *Zen-Brain Reflections: Reviewing Recent Developments in Meditation and States of Consciousness.* Cambridge, MA: MIT Press, 2006.

Avant, Lloyd L. "Vision in the Ganzfeld." *Psychological Bulletin* 64 (1965): 246–58.

Baden-Powell, Robert. *My Adventures as a Spy.* Cirencester, UK: The Echo Library, 2005 [1915].

Bakhtiarov, Oleg. "Psychonetics." August 13, 2009. http://www.psychotechnology.ru/publication/item56.html (accessed on July 13, 2012).

Bankier, David. *The Germans and the Final Solution: Public Opinion under Nazism.* Oxford: Blackwell, 1992.

Bartlett, Frederic C. *Remembering: A Study in Experimental and Social Psychology.* Cambridge: Cambridge University Press, 1964 [1932].

Bateson, Gregory. "A Theory of Play and Fantasy." In *Steps to an Ecology of Mind*, 177–93. New York: Ballantine Books, 1972 [1955].

Bateson, Mary C. *Peripheral Visions: Learning along the Way.* New York: HarperCollins, 1994.

Baumann, James F. "The Direct Instruction of Main Idea Comprehension Ability." In *Teaching Main Idea Comprehension*, 133–78. Newark, DE: International Reading Association, 1986.

——— (ed.). *Teaching Main Idea Comprehension.* Newark, DE: International Reading Association, 1986.

Baylis, Gordon C., and Ellison M. Cale. "The Figure Has a Shape, but the Ground Does Not: Evidence from a Priming Paradigm." *Journal of Experimental Psychology: Human Perception and Performance* 27 (2001): 633–43.

Bazerman, Max H., and Dolly Chugh. "Bounded Awareness: Focusing Failures in Negotiation." In Leigh L. Thompson (ed.), *Negotiation Theory and Research*, 7–26. New York: Psychology Press, 2006.

Becklen, Robert, and Daniel Cervone. "Selective Looking and the Noticing of Unexpected Events." *Memory and Cognition* 11 (1983): 601–08.

Beddard, Frank E. *Animal Coloration: An Account of the Principal Facts and Theories Relating to the Colours and Markings of Animals.* London: Swan Sonnenschein, 1895.

Behrens, Roy R. *Camoupedia: A Compendium of Research on Art, Architecture, and Camouflage.* Dysart, IA: Bobolink Books, 2009.

———. *False Colors: Art, Design, and Modern Camouflage.* Dysart, IA: Bobolink Books, 2002.

——— "Revisiting Gottschaldt: Embedded Figures in Art, Architecture, and Design." *Gestalt Theory: Journal of the GTA* 22, no. 2 (2000): 97–106.

——— (ed.). *Ship Shape: A Dazzle Camouflage Sourcebook.* Dysart, IA: Bobolink Books, 2012.

Berger, Harris M. *Metal, Rock, and Jazz: Perception and the Phenomenology of Musical Experience.* Middletown, CT: Wesleyan University Press, 1999.

Berger, Peter L., and Thomas Luckmann. *The Social Construction of Reality: A Treatise in the Sociology of Knowledge.* Garden City, NY: Anchor Books, 1967 [1966].

Bergson, Henri. "'Phantasms of the Living' and 'Psychical Research.'" In *Mind-Energy: Lectures and Essays,* 75–103. New York: Henry Holt, 1920 [1913].

Berlyne, D. E. *Aesthetics and Psychobiology.* New York: Appleton-Century-Crofts, 1971.

Berry, John W. "Ecological and Cultural Factors in Spatial Perceptual Development." *Canadian Journal of Behavioural Science* 3 (1971): 324–36.

———. *Human Ecology and Cognitive Style: Comparative Studies in Cultural and Psychological Adaptation.* New York: SAGE Publications, 1976.

———. "Temne and Eskimo Perceptual Skills." *International Journal of Psychology* 1 (1966): 207–29.

Bieri, James, et al. "Sex Differences in Perceptual Behavior." *Journal of Personality* 26 (1958): 1–12.

Bigand, E., et al. "Divided Attention in Music." *International Journal of Psychology* 35 (2000): 270–78.

Binet, Alfred, and Charles Féré. *Animal Magnetism.* 5th ed. London: Kegan Paul, Trench, Trübner, and Co., 1905 [1887].

Blake, William. *The Marriage of Heaven and Hell.* London: Oxford University Press, 1975 [1794].

Blechman, Hardy. *Disruptive Pattern Manual: An Encyclopedia of Camouflage.* Buffalo, NY: Firefly Books, 2004.

Bleuler, Eugen. *Dementia Praecox or the Group of Schizophrenias.* Madison, CT: International Universities Press, 1950 [1911].
Boduroglu, Aysecan, et al. "Cultural Differences in Allocation of Attention in Visual Information Processing." *Journal of Cross-Cultural Psychology* 40 (2009): 349–60.
Bogdashina, Olga. *Sensory Perceptual Issues in Autism and Asperger Syndrome: Different Sensory Experiences—Different Perceptual Worlds.* London: Jessica Kingsley Publishers, 2003.
Bothwell, Dorr, and Marlys Mayfield. *Notan: The Dark-Light Principle of Design.* New York: Dover Publications, 1991 [1968].
Bourdieu, Pierre. *Distinction: A Social Critique of the Judgement of Taste.* Cambridge, MA: Harvard University Press, 1984 [1979].
Bregman, Albert S. *Auditory Scene Analysis: The Perceptual Organization of Sound.* Cambridge, MA: MIT Press, 1990.
Bregman, Albert S., and Jeffrey Campbell. "Primary Auditory Stream Segregation and Perception of Order in Rapid Sequences of Tones." *Journal of Experimental Psychology* 89 (1971): 244–49.
Breitmeyer, Bruno G. *Blindspots: The Many Ways We Cannot See.* New York: Oxford University Press, 2010.
Brekhus, Wayne H. *Peacocks, Chameleons, Centaurs: Gay Suburbia and the Grammar of Social Identity.* Chicago: University of Chicago Press, 2003.
———. "Social Marking and the Mental Coloring of Identity: Sexual Identity Construction and Maintenance in the United States." *Sociological Forum* 11 (1996): 497–522.
———. "A Sociology of the Unmarked: Redirecting Our Focus." *Sociological Theory* 16 (1998): 34–51.
Bressan, Paola, and Silvia Pizzighello. "The Attentional Cost of Inattentional Blindness." *Cognition* 106 (2008): 370–83.
Broadbent, Donald E. *Perception and Communication.* Oxford: Pergamon Press, 1958.
Brochard, Renaud, et al. "Perceptual Organization of Complex Auditory Sequences: Effect of Number of Simultaneous Subsequences and Frequency Separation." *Journal of Experimental Psychology: Human Perception and Performance* 25 (1999): 1742–59.
Brown, Thomas E. *Attention Deficit Disorder: The Unfocused Mind in Children and Adults.* New Haven, CT: Yale University Press, 2005.
Brown, Tom. "Fill Your Senses, Light Up Your Life." *Reader's Digest*, August 1984, 153–56.

———. *Tom Brown's Field Guide to Nature Observation and Tracking.* New York: Berkley Books, 1983.

Brunius, Teddy. "Inside and Outside the Frame of a Work of Art." In *Idea and Form: Studies in the History of Art*, 1–23. Stockholm: Almqvist and Wiksell (Uppsala Studies in the History of Art, Figura Nova Series, no.1), 1959.

Burch, Bridgette. "The Misdirection in Attention." Unpublished paper, Rutgers University, New Brunswick, NJ, 2012.

Burkan, Wayne C. *Wide-Angle Vision: Beat Your Competition by Focusing on Fringe Competitors, Lost Customers, and Rogue Employees.* New York: John Wiley & Sons, 1996.

Buswell, Guy T. *How People Look at Pictures: A Study of the Psychology of Perception in Art.* Chicago: University of Chicago Press, 1935.

Butor, Michel. "Mondrian: The Square and Its Inhabitant." In *Inventory: Essays*, 235–52. New York: Simon and Schuster, 1968 [1965].

Cahill, Spencer E., et al. "Meanwhile Backstage: Public Bathrooms and the Interaction Order." *Urban Life* 14 (1985): 33–58.

Campion, Lisa. "The Social Construction of Attention: Varying Patterns of Attending, Inattending, and Disattending within Healthy and Sick Communities." Unpublished manuscript, Rutgers University, New Brunswick, NJ, 2013.

Carson, Shelley H., et al. "Decreased Latent Inhibition Is Associated with Increased Creative Achievement in High-Functioning Individuals." *Journal of Personality and Social Psychology* 85 (2003): 499–506.

Casati, Roberto, and Achille C. Varzi. *Holes and Other Superficialities.* Cambridge, MA: MIT Press, 1994.

Casson, Lionel. *Ships and Seamanship in the Ancient World.* Baltimore, MD: Johns Hopkins University Press, 1995 [1971].

Cavanagh, Patrick, and George A. Alvarez. "Tracking Multiple Targets with Multifocal Attention." *Trends in Cognitive Sciences* 9 (2005): 349–54.

Cave, Kyle R., and Narcisse P. Bichot. "Visuospatial Attention: Beyond a Spotlight Model." *Psychonomic Bulletin and Review* 6 (1999): 204–23.

Cerulo, Karen A. *Never Saw It Coming: Cultural Challenges to Envisioning the Worst.* Chicago: University of Chicago Press, 2006.

Chabris, Christopher, and Daniel Simons. *The Invisible Gorilla and Other Ways Our Intuition Deceives Us*. London: HarperCollins, 2010.

Chambers, Ross. "The Unexamined." In Mike Hill (ed.), *Whiteness: A Critical Reader*, 187–203. New York: New York University Press, 1997.

Chapman, Steve. *365 Things Every Hunter Should Know*. Eugene, OR: Harvest House Publishers, 2008.

Chavajay, Pablo, and Barbara Rogoff. "Cultural Variation in Management of Attention by Children and Their Caregivers." *Developmental Psychology* 35 (1999): 1079–90.

Cherry, E. Colin. "Some Experiments on the Recognition of Speech, with One and with Two Ears." *Journal of the Acoustical Society of America* 25 (1953): 975–79.

Chevan, David. "The Double Bass as a Solo Instrument in Early Jazz." *The Black Perspective in Music* 17 (1989): 73–91.

Chong, Sang C., and Karla K. Evans. "Distributed vs. Focused Attention (Count vs. Estimate)." *Wiley Interdisciplinary Reviews: Cognitive Science* 2 (2011): 634–38.

Chua, Hannah F., et al. "Cultural Variation in Eye Movements During Scene Perception." *Proceedings of the National Academy of Sciences of the United States of America* 102 (August 30, 2005): 12629–33.

Cobb, Roger W., and Charles D. Elder. "The Politics of Agenda-Building: An Alternative Perspective for Modern Democratic Theory." *Journal of Politics* 33 (1971): 892–915.

Cohen, Bernard C. *The Press and Foreign Policy*. Princeton, NJ: Princeton University Press, 1963.

Cohen, Stanley. *States of Denial: Knowing about Atrocities and Suffering*. Cambridge: Polity, 2001.

Cohen, Walter. "Form Recognition, Spatial Orientation, Perception of Movement in the Uniform Visual Field." In Ailene Morris and E. Porter Horne (eds.), *Visual Search Techniques*, 119–23. Washington, DC: Natural Academy of Sciences, 1960.

———. "Spatial and Textural Characteristics of the Ganzfeld." *American Journal of Psychology* 70 (1957): 403–10.

Collins, Allan, and Edward E. Smith. "Teaching the Process of Reading Comprehension." In Douglas K. Detterman and Robert J. Sternberg (eds.), *How and How Much Can Intelligence Be Increased*, 173–85. Norwood, NJ: Ablex, 1982.

Collins, Michael. "Smell Invisible: Being a Scent Free Hunter." *Cincinnati Hunting Examiner*, September 12, 2011. http://www.examiner.com/hunting-in-cincinnati/smell-invisible-being-a-scent-free-hunter.

Copland, Aaron. *What to Listen for in Music*. Rev. ed. New York: McGraw-Hill, 1957 [1939].

Correa-Chávez, Maricela, et al. "Cultural Patterns in Attending to Two Events at Once." *Child Development* 76 (2005): 664–78.

Cott, Hugh B. *Adaptive Coloration in Animals*. London: Methuen, 1957 [1940].

Crary, Jonathan. *Suspensions of Perception: Attention, Spectacle, and Modern Culture*. Cambridge, MA: MIT Press, 1999.

Crundall, David, et al. "Driving Experience and the Functional Field of View." *Perception* 28 (1999): 1075–87.

Cummings, Edward E. *A Selection of Poems*. New York: Harvest, 1965.

Cunningham, James W., and David W. Moore. "The Confused World of Main Idea." In James F. Baumann (ed.), *Teaching Main Idea Comprehension*, 1–17. Newark, DE: International Reading Association, 1986.

Cuthill, Innes C., and Tom S. Troscianko. "Animal Camouflage: Biology Meets Psychology, Computer Science, and Art." In C. A. Brebbia et al. (eds.), *Colour in Art, Design, and Nature*, 5–24. Ashurst, UK: WIT Press, 2011.

Cuthill, Innes C., et al. "Disruptive Coloration and Background Pattern Matching." *Nature* 434 (March 3, 2005): 72–74.

Darwin, Erasmus. *Zoonomia, or the Laws of Organic Life*. Vol. 1. London: J. Johnson, 1794.

Davenport, Thomas H., and John C. Beck. *The Attention Economy: Understanding the New Currency of Business*. Boston: Harvard Business School Press, 2001.

Davey, Beth. "Think Aloud: Modeling the Cognitive Processes of Reading Comprehension." *Journal of Reading* 27 (1983): 44–47.

Davis, Murray S. *Smut: Erotic Reality/Obscene Ideology*. Chicago: University of Chicago Press, 1983.

Day, George S., and Paul J. H. Schoemaker. *Peripheral Vision: Detecting the Weak Signals That Will Make or Break Your Company*. Boston: Harvard Business School Press, 2006.

Deikman, Arthur J. "Bimodal Consciousness." *Archives of General Psychiatry* 25 (1971): 481–89.

Deikman, Arthur J. "De-Automatization and the Mystic Experience." *Psychiatry* 29 (1966): 324–38.

Dellas, Marie, and Eugene L. Gaier. "Identification of Creativity: The Individual." *Psychological Bulletin* 73 (1970): 55–73.

Dennett, Daniel C. *Consciousness Explained*. Boston: Little, Brown & Co., 1991.

Denton, Jeremiah A., and Edwin H. Brandt. *When Hell Was in Session*. Traditional Press, 1982.

Derber, Charles. *The Pursuit of Attention: Power and Ego in Everyday Life*. 2nd ed. New York: Oxford University Press, 2000.

DesAutels, Peggy. "Gestalt Shifts in Moral Perception." In Larry May et al. (eds.), *Mind and Morals: Essays on Cognitive Science and Ethics*, 129–43. Cambridge, MA: MIT Press, 1996.

DeVault, Marjorie L. "Producing Family Time: Practices of Leisure Activity beyond the Home." *Qualitative Sociology* 23 (2000): 485–503.

Dewey, John. *Psychology*. 3rd ed. New York: Harper and Brothers, 1893.

Dilthey, Wilhelm. *Selected Works. Vol. I, Introduction to the Human Sciences*. Princeton, NJ: Princeton University Press, 1989 [1883].

Divenyi, Pierre L., and Ira J. Hirsh. "Some Figural Properties of Auditory Patterns." *Journal of the Acoustical Society of America* 64 (1978): 1369–85.

Doherty, Martin J., et al. "The Context-Sensitivity of Visual Size Perception Varies across Cultures." *Perception* 37 (2008): 1426–33.

Downs, Anthony. "Up and Down with Ecology: The 'Issue-Attention Cycle.'" *The Public Interest* 28 (1972): 38–50.

Doyle, Arthur C. "Silver Blaze." In *Sherlock Holmes: The Complete Novels and Stories*. Vol. 1, 521–46. New York: Bantam Books, 1986 [1892].

Driver, Jon, and Gordon C. Baylis. "Edge-Assignment and Figure-Ground Segmentation in Short-Term Visual Matching." *Cognitive Psychology* 31 (1996): 248–306.

Duncan, Andrew. *Centre and Periphery in Modern British Poetry*. Liverpool, UK: Liverpool University Press, 2005.

Duncan, John, and Glyn W. Humphreys. "Visual Search and Stimulus Similarity." *Psychological Review* 96 (1989): 433–58.

Dunn, Bruce R., et al. "Concentration and Mindfulness Meditations: Unique Forms of Consciousness?" *Applied Psychophysiology and Biofeedback* 24 (1999): 147–65.

Durkheim, Emile. *The Elementary Forms of Religious Life*. New York: Free Press, 1995 [1912].
———. "Individual and Collective Representations." In *Sociology and Philosophy*, 1–34. New York: Free Press, 1974 [1898].
Dykes, Margaret, and Andrew McGhie. "A Comparative Study of Attentional Strategies of Schizophrenic and Highly Creative Normal Subjects." *British Journal of Psychiatry* 128 (1976): 50–56.
Edwards, Betty. *Drawing on the Right Side of the Brain: A Course in Enhancing Creativity and Artistic Confidence*. Los Angeles: J. P. Tarcher, 1979.
———. *The New Drawing on the Right Side of the Brain*. New York: Jeremy P. Tarcher/Putnam, 1999.
Ehrenzweig, Anton. *The Hidden Order of Art: A Study in the Psychology of Artistic Imagination*. Berkeley: University of California Press, 1971 [1967].
———. *The Psycho-Analysis of Artistic Vision and Hearing: An Introduction to a Theory of Unconscious Perception*. New York: The Julian Press, 1953.
Ekman, Paul, and Wallace V. Friesen. "Nonverbal Leakage and Clues to Deception." *Psychiatry* 32 (1969): 88–106.
Ellison, Ralph. *Invisible Man*. New York: Random House, 1952.
Emerson, Joan P. "Behavior in Private Places: Sustaining Definitions of Reality in Gynecological Examinations." In Hans-Peter Dreitzel (ed.), *Recent Sociology No.2: Patterns of Communicative Behavior*, 74–97. New York: Macmillan, 1970.
Endler, John A. "A Predator's View of Animal Color Patterns." *Evolutionary Biology* 11 (1978): 319–64.
Escher, Maurits C. *Escher on Escher: Exploring the Infinite*. New York: Harry N. Abrams, 1989 [1986].
Fehmi, Les, and Jim Robbins. *The Open-Focus Brain: Harnessing the Power of Attention to Heal Mind and Body*. Boston: Trumpeter, 2008.
Felman, Shoshana, and Dori Laub (eds.). *Testimony: Crises of Witnessing in Literature, Psychoanalysis, and History*. New York: Routledge, 1992.
Fernald, Anne, and Hiromi Morikawa. "Common Themes and Cultural Variations in Japanese and American Mothers' Speech to Infants." *Child Development* 64 (1993): 637–56.

Fernandez-Duque, Diego, and Mark Johnson. "Attention Metaphors: How Metaphors Guide the Cognitive Psychology of Attention." *Cognitive Science* 23 (1999): 83–116.
Fine, Gary A. *Morel Tales: The Culture of Mushrooming.* Cambridge, MA: Harvard University Press, 1998.
Fingarette, Herbert. *Self-Deception.* London: Routledge and Kegan Paul, 1969.
Fisher, Charles. "Further Observations on the Poetzl Phenomenon: The Effects of Subliminal Visual Stimulation on Dreams, Images, and Hallucinations." *Psychoanalysis and Contemporary Thought* 11 (1988): 3–56.
Fleck, Ludwik. *Genesis and Development of A Scientific Fact.* Chicago: University of Chicago Press, 1979 [1935].
Fleischman, Suzanne. "Discourse Functions of Tense-Aspect Oppositions in Narrative: Toward a Theory of Grounding." *Linguistics* 23 (1985): 851–82.
Follette, Victoria M., et al. "Acceptance, Mindfulness, and Trauma." In Steven C. Hayes et al. (eds.), *Mindfulness and Acceptance: Expanding the Cognitive-Behavioral Tradition*, 192–208. New York: Guilford, 2004.
Force, William R. "The Code of Harry: Performing Normativity in Dexter." *Crime, Media, Culture* 6 (2010): 329–45.
Foucault, Michel. *The Birth of the Clinic: An Archaeology of Medical Perception.* New York: Vintage Books, 1975 [1963].
Francolini, Carl M., and Howard E. Egeth. "Perceptual Selectivity Is Task Dependent: The Pop-Out Effect Poops Out." *Perception and Psychophysics* 25 (1979): 99–110.
Freeman, Thomas, et al. *Studies on Psychosis: Descriptive, Psychoanalytic, and Psychological Aspects.* New York: International Universities Press, 1966.
Freud, Anna. *The Ego and the Mechanisms of Defence.* London: Karnak Books, 1993 [1936].
Freud, Sigmund. *The Psychopathology of Everyday Life.* New York: W. W. Norton, 1960 [1901].
Freudenburg, William R., and Margarita Alario. "Weapons of Mass Distraction: Magicianship, Misdirection, and the Dark Side of Legitimation." *Sociological Forum* 22 (2007): 146–73.
Freyd, Jennifer J. *Betrayal Trauma: The Logic of Forgetting Childhood Abuse.* Cambridge, MA: Harvard University Press, 1996.

Freyd, Jennifer J., and Pamela Birrell. *Blind to Betrayal: Why We Fool Ourselves We Aren't Being Fooled*. Hoboken, NJ: John Wiley & Sons, 2013.

Friedman, Asia. *Blind to Sameness: Sexpectations and the Social Construction of Male and Female Bodies*. Chicago: University of Chicago Press, 2013.

———. "Toward a Sociology of Perception: Sight, Sex, and Gender." *Cultural Sociology* 5 (2011): 187–206.

Friedman, Ronald S., et al. "Attentional Priming Effects on Creativity." *Creativity Research Journal* 15 (2003): 277–86.

Friend, Trudy. *Landscape: Problems and Solutions*. Newton Abbot, UK: David & Charles, 2004.

Frith, Uta, and Francesca Happé. "Autism: Beyond 'Theory of Mind.'" *Cognition* 50 (1994): 115–32.

Gans, Herbert J. *Deciding What's News: A Study of CBS Evening News, NBC Nightly News, Newsweek, and Time*. New York: Random House, 1979.

Garfinkel, Harold. "Studies of the Routine Grounds of Everyday Activities." In *Studies in Ethnomethodology*, 35–75. Englewood Cliffs, NJ: Prentice-Hall, 1967 [1964].

Garland-Thomson, Rosemarie. *Staring: How We Look*. New York: Oxford University Press, 2009.

Ghent, Lila. "Perception of Overlapping Figures by Children of Different Ages." *American Journal of Psychology* 69 (1956): 575–87.

Gibson, James J. "Perception of Distance and Space in the Open Air." In David C. Beardslee and Michael Wertheimer (eds.), *Readings in Perception*, 415–31. Princeton, NJ: D. Van Nostrand Company, 1958 [1946].

———. *The Perception of the Visual World*. Westport, CT: Greenwood Press, 1950.

———. *The Senses Considered as Perceptual Systems*. Boston: Houghton Mifflin, 1966.

Giedion, Sigfried. *Space, Time, and Architecture: The Growth of A New Tradition*. 3rd ed. Cambridge, MA: Harvard University Press, 1956.

Gilligan, Carol. "Moral Orientation and Moral Development." In Eva F. Kittay and Diana T. Meyers (eds.), *Women and Moral Theory*, 19–33. Totowa, NJ: Rowman & Littlefield, 1987.

Goffman, Erving. "Alienation from Interaction." In *Interaction Ritual: Essays on Face-to-Face Behavior*, 113–36. Garden City, NY: Anchor Books, 1967 [1957].
———. *Behavior in Public Places: Notes on the Social Organization of Gatherings*. New York: Free Press, 1963.
———. "Footing." In *Forms of Talk*, 124–59. Philadelphia: University of Pennsylvania Press, 1981 [1979].
———. *Frame Analysis: An Essay on the Organization of Experience*. New York: Harper & Row, 1974.
———. "Fun in Games." In *Encounters: Two Studies in the Sociology of Interaction*, 15–81. Indianapolis: Bobbs-Merrill, 1961.
———. "On Face Work: An Analysis of Ritual Elements in Social Interaction." In *Interaction Ritual: Essays on Face-to-Face Behavior*, 5–45. Garden City, NY: Anchor Books, 1967 [1955].
———. *Relations in Public: Microstudies of the Public Order*. New York: Harper & Row, 1971.
———. *The Presentation of Self in Everyday Life*. Garden City, NY: Doubleday Anchor, 1959.
Gold, Steven N., and Gonzalo Bacigalupe. "Interpersonal and Systemic Theories of Personality." In David F. Barone et al. (eds.), *Advanced Personality*, 57–79. New York: Springer, 1998.
Gombrich, Ernst H. *Art and Illusion: A Study in the Psychology of Pictorial Representation*. London: Phaidon, 2002 [1960].
Goodwin, Charles. "Professional Vision." *American Anthropologist* 96 (1994): 606–33.
Goodwin, Charles, and Alessandro Duranti. "Rethinking Context: An Introduction." In Alessandro Duranti and Charles Goodwin (eds.), *Rethinking Context: Language as an Interactive Phenomenon*, 1–42. Cambridge: Cambridge University Press, 1992.
Gottschaldt, Kurt. "The Influence of Past Experience on the Perception of Figures." In M. D. Vernon (ed.), *Experiments in Visual Perception: Selected Readings*, 29–44. Harmondsworth, UK: Penguin Books, 1966 [1926].
Gottsdanker, Robert. "The Relation between the Nature of the Search Situation and the Effectiveness of Alternative Strategies of Search." In Ailene Morris and E. Porter Horne (eds.), *Visual Search Techniques*, 181–86. Washington, DC: Natural Academy of Sciences, 1960.

Grandin, Temple, and Catherine Johnson. *Animals in Translation: Using the Mysteries of Autism to Decode Animal Behavior.* Orlando, FL: Harvest Books, 2006 [2005].

Grandin, Temple, and Richard Panek. *The Autistic Brain: Thinking Across the Spectrum.* Boston: Houghton Mifflin Harcourt, 2013.

Grauds, Constance, and Doug Childers. *The Energy Prescription: Give Yourself Abundant Vitality with the Wisdom of America's Leading Natural Pharmacist.* New York: Bantam Books, 2005.

Grazian, David. "Some Animals Are More Equal Than Others: American Zoos and the Culture of Childhood." Paper presented at the annual meeting of the American Sociological Association, New York, August 2013.

Graziano, Frank. *Divine Violence: Spectacle, Psychosexuality, and Radical Christianity in the Argentine "Dirty War."* Boulder, CO: Westview, 1992.

Green, John. *Ballet Class Coloring Book.* Mineola, NY: Dover, 1998.

———. *The Language of Flowers Coloring Book.* Mineola, NY: Dover, 2003.

Gregory, Andrew H. "Listening to Polyphonic Music." *Psychology of Music* 18 (1990): 163–70.

Griffiths, Tom, and Cathleen Moore. "A Matter of Perception." *Aquatics International*, November/December 2004. http://www.aquaticsintl.com/2004/nov/0411_rm.html (accessed on August 30, 2012).

Grossman, Valerie G. A. *Quick Reference to Triage.* 2nd ed. Philadelphia, PA: Lippincott Williams & Wilkins, 2003.

Gugerty, Leo. "Situation Awareness in Driving." In Donald L. Fisher et al. (eds.), *Handbook of Driving Simulation for Engineering, Medicine, and Psychology*, 19.265–19.272. Boca Raton, FL: CRC Press, 2011.

Gurwitsch, Aron. *The Field of Consciousness.* Pittsburgh, PA: Duquesne University Press, 1964.

Gutchess, Angela H., et al. "Cultural Differences in Neural Function Associated with Object Processing." *Cognitive, Affective, and Behavioral Neuroscience* 6 (2006): 102–09.

Haaken, Janice. "Field Dependence Research: A Historical Analysis of a Psychological Construct." *Signs* 13 (1988): 311–30.

Hall, Edward T. *Beyond Culture.* Garden City, NY: Anchor Books, 1977 [1976].

Hall, Edward T. *The Hidden Dimension*. Garden City, NY: Doubleday, 1966.
Halley, Richard D. "Distractability of Males and Females in Competing Aural Message Situations: A Research Note." *Human Communication Research* 2 (1975): 79–82.
Halton, Eugene. "Peircean Animism and the End of Civilization." *Contemporary Pragmatism* 2 (2005): 135–66.
Handford, Martin. *Where's Waldo?* Somerville, MA: Candlewick Press, 2007 [1987].
Happé, Francesca. *Autism: An Introduction to Psychological Theory*. Cambridge, MA: Harvard University Press, 1995.
———. "Autism: Cognitive Deficit or Cognitive Style?" *Trends in Cognitive Sciences* 3 (1999): 216–22.
Harrington, Jonathan, and Steve Cassidy. *Techniques in Speech Acoustics*. Dordrecht, The Netherlands: Kluwer Academic Publishers, 1999.
Hartmann, Thom. *Attention Deficit Disorder: A Different Perception*. Grass Valley, CA: Underwood Books, 1997 [1993].
Hayes, Steven C., et al. "Experiential Avoidance and Behavioral Disorders: A Functional Dimensional Approach to Diagnosis and Treatment." *Journal of Consulting and Clinical Psychology* 64 (1996): 1152–68.
Heaton, Pamela F. "Pitch Memory, Labelling, and Disembedding in Autism." *Journal of Child Psychology and Psychiatry* 44 (2003): 543–51.
Heilman, Samuel C. *A Walker in Jerusalem*. New York: Summit Books, 1986.
Heller, Joseph. *We Bombed in New Haven*. New York: Delta, 1967.
Helson, Harry. "The Psychology of Gestalt." *American Journal of Psychology* 36 (1925): 494–526.
Hernández-Peón, Raúl. "Psychiatric Implications of Neurophysiological Research." *Bulletin of the Menninger Clinic* 28 (1964): 165–84.
Hilgartner, Stephen, and Charles L. Bosk. "The Rise and Fall of Social Problems: A Public Arenas Model." *American Journal of Sociology* 94 (1988): 53–78.
Hirschauer, Stefan. "The Manufacture of Bodies in Surgery." *Social Studies of Science* 21 (1991): 279–319.
Hochberg, Julian E. *Perception*. Englewood Cliffs, NJ: Prentice-Hall, 1964.

Hochschild, Arlie. *The Managed Heart: Commercialization of Human Feeling.* Berkeley: University of California Press, 1983.

Hodges, Donald A., and David C. Sebald. *Music in the Human Experience: An Introduction to Music Psychology.* New York: Routledge, 2011.

Hofstadter, Douglas R. "Changes in Default Words and Images, Engendered by Rising Consciousness." In *Metamagical Themas: Questing for the Essence of Mind and Pattern,* 136–58. New York: Basic Books, 1985 [1982].

———. *Gödel, Escher, Bach: An Eternal Golden Braid.* New York: Basic Books, 1979.

Hopper, Paul J., and Sandra A. Thompson. "Transitivity in Grammar and Discourse." *Language* 56 (1980): 251–99.

Horn, David, and Stan Hawkins. "Backing." In John Shepherd et al. (eds.), *Continuum Encyclopedia of Popular Music of the World.* Vol. II, *Performance and Production,* 632. London: Continuum, 2003.

Horowitz, Alexandra. *On Looking: Eleven Walks with Expert Eyes.* New York: Scribner, 2013.

Horowitz, Seth S. "The Science and Art of Listening." *New York Times,* November 11, 2012. http://www.nytimes.com/2012/11/11/opinion/sunday/why-listening-is-so-much-more-than-hearing.html?_r=0 (accessed April 22, 2013).

Horwitz, Allan V. "Normality." *Contexts* 7, no. 1 (2008): 70–71.

Horwitz, Gordon J. *In the Shadow of Death: Living Outside the Gates of Mauthausen.* New York: Free Press, 1990.

Hotchkiss, John. "Children and Conduct in a Ladino Community in Chiapas, Mexico." *American Anthropologist* 69 (1967): 711–18.

Howell, Elizabeth F. *The Dissociative Mind.* New York: Routledge, 2005.

Hsiao, Steven S., and Francisco Vega-Bermudez. "Attention in the Somatosensory System." In Randall J. Nelson (ed.), *The Somatosensory System: Deciphering the Brain's Own Body Image,* 197–217. Boca Raton, FL: CRC Press LLC, 2001.

Hubert, Henri. "A Brief Study of the Representation of Time in Religion and Magic." In *Essay on Time,* 43–91. Oxford: Durkheim Press, 1999 [1905].

Huff, Richard, and Brian Kates. "TV to Screen Bin Laden Tapes: White House Warns They May Contain Coded Messages." *New*

York Daily News, October 11, 2001. http://articles.nydailynews.com/2001-10-11/news/18369890_1_ laden-bin-messages.

Humphreys, Glyn W., et al. "Attending but Not Seeing: The 'Other Race' Effect in Face and Person Perception Studied through Change Blindness." *Visual Cognition* 12 (2005): 249–62.

Huron, David. "Tonal Consonance versus Tonal Fusion in Polyphonic Sonorities." *Music Perception* 9 (1991): 135–54.

———. "Voice Denumerability in Polyphonic Music of Homogeneous Timbres." *Music Perception* 6 (1989): 361–82.

Husserl, Edmund. *Thing and Space: Lectures of 1907*. Dordrecht, The Netherlands: Kluwer Academic Publishers, 2010 [1907].

Huxley, Aldous. "The Doors of Perception." In *The Doors of Perception and Heaven and Hell*, 9–79. New York: HarperCollins, 2009 [1954].

Iserson, Kenneth V., and John C. Moskop. "Triage in Medicine, Part I: Concept, History, and Types." *Annals of Emergency Medicine* 49 (2007): 275–81.

Iwasaki, Syoichi. "Spatial Attention and Two Modes of Visual Consciousness." *Cognition* 49 (1993): 211–33.

Iyengar, Shanto, and Donald R. Kinder. *News That Matters: Television and American Opinion*. Chicago: University of Chicago Press, 2010 [1987].

Jaarsma, Pier, and Stellan Welin. "Autism as a Natural Human Variation: Reflections on the Claims of the Neurodiversity Movement." *Health Care Analysis* February 11, 2011. http://www.imh.liu.se/avd_halsa_samhalle/filarkiv1/1.264263/JaarsmaWelin2011Autismasanaturalvariation.pdf.

James, William. *The Principles of Psychology*. Cambridge, MA: Harvard University Press, 1983 [1890].

Jastrow, Joseph. *The Subconscious*. Boston: Houghton, Mifflin & Co., 1906.

Ji, Li-Jun, et al. "Culture, Control, and Perception of Relationships in the Environment." *Journal of Personality and Social Psychology* 78 (2000): 943–55.

Jolliffe, Therese, and Simon Baron-Cohen. "Are People with Autism and Asperger Syndrome Faster Than Normal on the Embedded Figures Test?" *Journal of Child Psychology and Psychiatry* 38 (1997): 527–34.

Jones, Bryan D., and Frank R. Baumgartner. *The Politics of Attention: How Government Prioritizes Problems*. Chicago: University of Chicago Press, 2005.

Jones, Dave. *Basketball—It's All about the Shot*. http://www.basketballshootingcoach.com/files/1625141/uploaded/BBShootingBookPDF2.pdf.

Jones, Warren, and Ami Klin. "Attention to Eyes Is Present but in Decline in 2–6-Month-Old Infants Later Diagnosed with Autism." *Nature* 12715 (November 6, 2013).

Joseph, Eddie. *How to Pick Pockets for Fun and Profit: A Magician's Guide to Pickpocket Magic*. Colorado Springs, CO: Piccadilly Books, 1992.

Julesz, Bela. "Perceptual Limits of Texture Discrimination and Their Implications to Figure-Ground Separation." In Emanuel L. J. Leeuwenberg and H. F. J. M. Buffart (eds.), *Formal Theories of Visual Perception*, 205–16. New York: John Wiley & Sons, 1978.

Jung, Richard. "Correlation of Bioelectrical and Autonomic Phenomena with Alterations of Consciousness and Arousal in Man." In J. F. Delafresnaye (ed.), *Brain Mechanisms and Consciousness*, 310–44. Springfield, IL: Charles C. Thomas, 1954.

Kabat-Zinn, Jon. *Full Catastrophe Living: Using the Wisdom of Your Body and Mind to Face Stress, Pain, and Illness*. New York: Delacorte Press, 1990.

Kahneman, Daniel. *Attention and Effort*. Englewood Cliffs, NJ: Prentice-Hall, 1973.

Kanizsa, Gaetano. "Perception, Past Experience, and the 'Impossible Experiment.'" In *Organization in Vision: Essays on Gestalt Perception*, 25–54. New York: Praeger, 1979 [1969].

Kaprow, Allan. *Assemblage, Environments, and Happenings*. New York: Harry N. Abrams, 1966.

Kashatus, William C. *Just Over the Line: Chester County and the Underground Railroad*. West Chester, PA: Chester County Historical Society, 2002.

Kasof, Joseph. "Creativity and Breadth of Attention." *Creativity Research Journal* 10 (1997): 303–15.

Katz, David. *The World of Touch*. Hillsdale, NJ: Lawrence Erlbaum Associates, 1989 [1925].

Keefe, Thomas. "Meditation and Social Work Treatment." In Francis J. Turner (ed.), *Social Work Treatment: Interlocking Theoretical Approaches*, 434–60. 4th ed. New York: Free Press, 1996.

Keen, Angeline M. "Protective Coloration in the Light of Gestalt Theory." *Journal of General Psychology* 6 (1932): 200–03.

Keller, Heidi. *Cultures of Infancy*. Mahwah, NJ: Lawrence Erlbaum Associates, 2007.

Kelner, Shaul. *Tours That Bind: Diaspora, Pilgrimage, and Israeli Birthright Tourism*. New York: New York University Press, 2010.

Kendon, Adam. "The Negotiation of Context in Face-to-Face Interaction." In Alessandro Duranti and Charles Goodwin (eds.), *Rethinking Context: Language as an Interactive Phenomenon*, 323–34. Cambridge: Cambridge University Press, 1992.

Kepes, Gyorgy. *Language of Vision*. Chicago: Paul Theobald, 1951.

Kern, Stephen. "Cubism, Camouflage, Silence, and Democracy: A Phenomenological Approach." In Roger Friedland and Deirdre Boden (eds.), *NowHere: Space, Time, and Modernity*, 163–80. Berkeley: University of California Press, 1994.

———. *The Culture of Time and Space 1880–1918*. Cambridge, MA: Harvard University Press, 1983.

Kingdon, John W. *Agendas, Alternatives, and Public Policies*. New York: HarperCollins, 1984.

Kinsbourne, Marcel, and Paula J. Caplan. *Children's Learning and Attention Problems*. Boston: Little, Brown, and Co., 1979.

Kirby, Michael (ed.). *Happenings: An Illustrated Anthology*. New York: E. P. Dutton, 1965.

Kirby, Michael, and Victoria N. Kirby. *Futurist Performance*. New York: PAJ Publications, 1986.

Kitayama, Shinobu, et al. "Perceiving an Object and Its Context in Different Cultures: A Cultural Look at New Look." *Psychological Science* 14 (2003): 201–06.

Kjolseth, Rolf. "Making Sense: Natural Language and Shared Knowledge in Understanding." In Joshua A. Fishman (ed.), *Advances in the Sociology of Language. Vol. II, Selected Studies and Applications*, 50–76. The Hague, The Netherlands: Mouton, 1972.

Klin, Ami, et al. "Visual Fixation Patterns During Viewing of Naturalistic Social Situations as Predictors of Social Competence in Individuals with Autism." *Archives of General Psychiatry* 59 (2002): 809–16.

Klinenberg, Eric. *Heat Wave: A Social Autopsy of Disaster in Chicago.* Chicago: University of Chicago Press, 2002.

Koffka, Kurt. "Perception: An Introduction to the *Gestalt-Theorie.*" *Psychological Bulletin* 19 (1922): 531–85.

———. *Principles of Gestalt Psychology.* New York: Harcourt, Brace & World, 1935.

Köhler, Wolfgang. *Gestalt Psychology: An Introduction to New Concepts in Modern Psychology.* New York: Mentor Books, 1947.

Kohn, Livia. "Meditation and Visualization." In Fabrizio Pregadio (ed.), *The Encyclopedia of Taoism.* Vol. 1, 118–20. New York: Routledge, 2008.

Kostelanetz, Richard. *The Theater of Mixed Means.* New York: Dial Press, 1968.

Kostelnick, Charles, and David D. Roberts. *Designing Visual Language: Strategies for Professional Communicators.* Boston: Allyn and Bacon, 1998.

Krupinski, Elizabeth A. "Visual Scanning Patterns of Radiologists Searching Mammograms." *Academic Radiology* 3 (1996): 137–44.

Kubovy, Michael, and David Van Valkenburg. "Auditory and Visual Objects." *Cognition* 80 (2001): 97–126.

Kuhn, Gustav, and John M. Findlay. "Misdirection, Attention, and Awareness: Inattentional Blindness Reveals Temporal Relationship between Eye Movements and Visual Awareness." *Quarterly Journal of Experimental Psychology* 63 (2010): 136–46.

Kuhn, Gustav, et al. "Towards a Science of Magic." *Trends in Cognitive Sciences* 12 (2008): 349–54.

Kuhn, Thomas S. *The Structure of Scientific Revolutions.* 2nd ed. Chicago: University of Chicago Press, 1970 [1962].

Kühnen, Ulrich, et al. "Cross-Cultural Variations in Identifying Embedded Figures: Comparisons from the United States, Germany, Russia, and Malaysia." *Journal of Cross-Cultural Psychology* 32 (2001): 366–72.

Kundel, Harold L., and Calvin F. Nodine. "Studies of Eye Movements and Visual Search in Radiology." In John W. Senders et al. (eds.), *Eye Movements and the Higher Psychological Functions,* 317–28. Hillsdale, NJ: Lawrence Erlbaum Associates, 1978.

Kuran, Timur. *Private Truths, Public Lies: The Social Consequences of Preference Falsification.* Cambridge, MA: Harvard University Press, 1995.

Kusakov, Igor. "Deconcentration of Attention: Addressing the Complexity of Software Engineering." January 27, 2012. http://deconcentration-of-attention.com/ (accessed on July 13, 2012).

LaBerge, David. *Attentional Processing: The Brain's Art of Mindfulness.* Cambridge, MA: Harvard University Press, 1995.

Lakoff, George, and Mark Johnson. *Metaphors We Live By.* Chicago: University of Chicago Press, 1980.

Lamont, Peter, and Richard Wiseman. *Magic in Theory: An Introduction to the Theoretical and Psychological Elements of Conjuring.* Hatfield, UK: University of Hertfordshire Press, 1999.

Lane, David M., and Deborah A. Pearson. "The Development of Selective Attention." *Merrill-Palmer Quarterly* 28 (1982): 317–37.

Lang, John T. "Sound and the City: Noise in Restaurant Critics' Reviews." Paper presented at the annual meeting of the American Sociological Association, New York, August 2013.

Lantigua-Williams, Juleyka. "Missing & Black: Helping Families Cope." http://jetmag.com/news/missing-black-helping-families-cope (accessed on May 29, 2013).

Lanza, Joseph. *Elevator Music: A Surreal History of Muzak, Easy-Listening, and Other Moodsong.* Rev. ed. Ann Arbor: University of Michigan Press, 2004.

———. "The Sound of Cottage Cheese (Why Background Music Is the Real World Beat!)" *Performing Arts Journal* 13, no. 3 (September 1991): 42–53.

Lanzmann, Claude. *Shoah: An Oral History of the Holocaust.* New York: Pantheon, 1985.

Laqueur, Thomas. *Making Sex: Body and Gender from the Greeks to Freud.* Cambridge, MA: Harvard University Press, 1990.

Leder, Drew. *The Absent Body.* Chicago: University of Chicago Press, 1990.

Lesgold, Alan, et al. "Expertise in a Complex Skill: Diagnosing X-Ray Pictures." In Michelene T. H. Chi et al. (eds.), *The Nature of Expertise*, 311–42. Hillsdale, NJ: Lawrence Erlbaum Associates, 1988.

Lewis, Richard S., et al. "Culture and Context: East Asian American and European American Differences in P3 Event-Related Potentials and Self-Construal." *Personality and Social Psychology Bulletin* 34 (2008): 623–34.

Linden, William. "Practicing of Meditation by School Children and Their Levels of Field Dependence-Independence, Test Anxiety, and Reading Achievement." *Journal of Consulting and Clinical Psychology* 41 (1973): 139–43.

Luchterhand, Elmer. "Knowing and Not Knowing: Involvement in Nazi Genocide." In Paul Thompson (ed.), *Our Common History: The Transformation of Europe*, 251–72. Atlantic Highlands, NJ: Humanities Press, 1982.

Luckiesh, Matthew. *Visual Illusions: Their Causes, Characteristics, and Applications*. New York: D. Van Nostrand & Constable, 1922.

Maccoby, Eleanor. "Sex Differences in Intellectual Functioning." In *The Development of Sex Differences*, 25–55. Palo Alto, CA: Stanford University Press, 1966.

MacDonald, Norma. "Living with Schizophrenia." *Canadian Medical Association Journal* 82 (1960): 218–21.

Mack, Arien. "Inattentional Blindness: Looking Without Seeing." *Current Directions in Psychological Science* 12 (2003): 180–84.

Mack, Arien, and Irvin Rock. *Inattentional Blindness*. Cambridge, MA: MIT Press, 1998.

Mack, Arien, et al. "Perceptual Organization and Attention." *Cognitive Psychology* 24 (1992): 475–501.

Macknik, Stephen L., et al. "Attention and Awareness in Stage Magic: Turning Tricks into Research." *Nature Reviews Neuroscience* 9 (2008): 871–79.

Mac Low, Jackson. *Representative Works 1938–1985*. New York: Roof, 1986.

Mallarmé, Stéphane. *A Tomb for Anatole*. San Francisco: North Point, 1983 [1961].

Markus, Hazel R., and Shinobu Kitayama. "Culture and the Self: Implications for Cognition, Emotion, and Motivation." *Psychological Review* 98 (1991): 224–53.

Martin, Judith. *Miss Manners' Guide for the Turn-of-the-Millennium*. New York: Fireside, 1990.

Mason-Schrock, Douglas. "Transsexuals' Narrative Construction of the 'True Self.'" *Social Psychology Quarterly* 59 (1996): 176–92.

Masuda, Takahiko, and Richard E. Nisbett. "Attending Holistically versus Analytically: Comparing the Context Sensitivity of Japanese and Americans." *Journal of Personality and Social Psychology* 81 (2001): 922–34.

———. "Culture and Change Blindness." *Cognitive Science* 30 (2006): 381–99.

Masuda, Takahiko, et al. "Culture and Aesthetic Preference: Comparing the Attention to Context of East Asians and Americans." *Personality and Social Psychology Bulletin* 34 (2008): 1260–75.

Mazza, Veronica, et al. "Foreground-Background Segmentation and Attention: A Change Blindness Study." *Psychological Research* 69 (2005): 201–10.

McAdams, Stephen, and Albert Bregman. "Hearing Musical Streams." *Computer Music Journal* 3, no. 4 (December 1979): 26–43, 60.

McCombs, Maxwell. *Setting the Agenda: The Mass Media and Public Opinion.* Cambridge, UK: Polity Press, 2004.

McCombs, Maxwell E., and Donald L. Shaw. "The Agenda-Setting Function of Mass Media." *Public Opinion Quarterly* 36 (1972): 176–87.

McGhie, Andrew. *Pathology of Attention.* Baltimore, MD: Penguin Books, 1969.

McGhie, Andrew, and James Chapman. "Disorders of Attention and Perception in Early Schizophrenia." *British Journal of Medical Psychology* 34 (1961): 103–16.

Memmert, Daniel. "The Effects of Eye Movements, Age, and Expertise on Inattentional Blindness." *Consciousness and Cognition* 15 (2006): 620–27.

Memmert, Daniel, and Philip Furley. "'I Spy with My Little Eye!': Breadth of Attention, Inattentional Blindness, and Tactical Decision Making in Team Sports." *Journal of Sport and Exercise Psychology* 29 (2007): 365–81.

Merilaita, Sami, and Johan Lind. "Background-Matching and Disruptive Coloration, and the Evolution of Cryptic Coloration." *Proceedings of the Royal Society B: Biological Sciences* 272 (2005): 665–70.

Merleau-Ponty, Maurice. *Phenomenology of Perception.* New York: Humanities Press, 1962 [1945].

Metzger, Wolfgang. *Laws of Seeing.* Cambridge, MA: MIT Press, 2006 [1936].

Meyrowitz, Joshua, "The Press Rejects a Candidate." *Columbia Journalism Review* 31 (March/April 1992): 46–47.

Miall, David S., and Don Kuiken. "Foregrounding, Defamiliarization, and Affect: Response to Literary Stories." *Poetics* 22 (1994): 389–407.

Miller, Leon K. *Musical Savants: Exceptional Skill in the Mentally Retarded*. Hillsdale, NJ: Lawrence Erlbaum, 1989.

Mills, C. Wright. "Methodological Consequences of the Sociology of Knowledge." *American Journal of Sociology* 46 (1940): 316–30.

Mills, Charles W. *The Racial Contract*. Ithaca, NY: Cornell University Press, 1997.

Milner, Marion. "The Framed Gap." In *The Suppressed Madness of Sane Men: Forty-Four Years of Exploring Psychoanalysis*, 79–82. London: Tavistock, 1987 [1952].

———. *A Life of One's Own*. London: Routledge, 2011 [1934].

Miner, Horace. "Body Ritual among the Nacirema." *American Anthropologist* 58 (1956): 503–07.

Minissale, Gregory. *The Psychology of Contemporary Art*. Cambridge, UK: Cambridge University Press, 2013.

Mishler, Elliot G. "Meaning in Context: Is There Any Other Kind?" *Harvard Educational Review* 49 (1979): 1–19.

Miyamoto, Yuri, et al. "Culture and the Physical Environment: Holistic versus Analytic Perceptual Affordances." *Psychological Science* 17 (2006): 113–19.

Moir, Anne, and David Jessel. *Brain Sex: The Real Difference Between Men and Women*. New York: Lyle Stuart, 1991.

Monahan, Sean. "Camouflaged Communication: Hiding a Connection between Signifier and Signified." Unpublished paper, Rutgers University, New Brunswick, NJ, 2002.

Moran, Aidan. "Attention in Sport." In Stephen D. Mellalieu and Sheldon Hanton (eds.), *Advances in Applied Sport Psychology: A Review*, 195–220. London: Routledge, 2009.

Mourant, Ronald R., and Thomas H. Rockwell. "Strategies of Visual Search by Novice and Experienced Drivers." *Human Factors* 14 (1972): 325–35.

Neisser, Ulric. *Cognitive Psychology*. New York: Appleton-Century-Crofts, 1967.

———. "Visual Search." *Scientific American* 210 (June 1964): 94–102.

Neisser, Ulric, and Robert Becklen. "Selective Looking: Attending to Visually Specified Events." *Cognitive Psychology* 7 (1975): 480–94.

Newark, Tim. *Camouflage*. London: Thames and Hudson, 2007.

Nisbett, Richard E. *The Geography of Thought: How Asians and Westerners Think Differently . . . and Why*. New York: Free Press, 2003.

Nisbett, Richard E., and Yuri Miyamoto. "The Influence of Culture: Holistic vs. Analytic Perception." *Trends in Cognitive Sciences* 9 (2005): 467–73.

Nodine, Calvin F., and Harold L. Kundel. "Using Eye Movements to Study Visual Search and to Improve Tumor Detection." *RadioGraphics* 7 (1987): 1241–50.

Nodine, Calvin F., and Claudia Mello-Thoms. "The Nature of Expertise in Radiology." In Jacob Beutel et al. (eds.), *Handbook of Medical Imaging.* Vol. 1, *Physics and Psychophysics*, 859–94. Bellingham, WA: SPIE Press, 2000.

Nodine, Calvin F., et al. "Eye Movements During Visual Search for Artistically Embedded Targets." *Bulletin of the Psychonomic Society* 13 (1979): 371–74.

Nodine, Calvin F., et al. "Searching for Nina." In John W. Senders et al. (eds.), *Eye Movements and the Higher Psychological Functions*, 241–58. Hillsdale, NJ: Lawrence Erlbaum Associates, 1978.

Nodine, Calvin F., et al. "The Role of Formal Art Training on Perception and Aesthetic Judgment of Art Compositions." *Leonardo* 26 (1993): 219–27.

Norgaard, Kari M. *Living in Denial: Climate Change, Emotions, and Everyday Life.* Cambridge, MA: MIT Press, 2011.

Nougier, Vincent, et al. "Information Processing in Sport and 'Orienting of Attention.'" *International Journal of Sport Psychology* 22 (1991): 307–27.

Oakley, Todd. *From Attention to Meaning: Explorations in Semiotics, Linguistics, and Rhetoric.* Bern, Switzerland: Peter Lang, 2009.

Ocasio, William. "Towards an Attention-Based View of the Firm." *Strategic Management Journal* 18 (1997): 187–206.

Ochs, Elinor. *Culture and Language Development: Language Acquisition and Language Socialization in a Samoan Village.* New York: Cambridge University Press, 1988.

Odets, Clifford. "Waiting for Lefty." In *Waiting for Lefty and Other Plays*, 5–31. New York: Grove Press, 1993 [1935].

Olsen, Mark. "'20 Feet from Stardom' Moves Spotlight to the Background." *Los Angeles Times*, June 18, 2013. http://articles.latimes.com/2013/jun/18/entertainment/la-et-mn-20-feet-from-stardom-20130619.

Ornstein, Robert. *Meditation and Modern Psychology.* Los Altos, CA: Malor Books, 2008 [1971].

Paris, Scott, et al. "Informed Strategies for Learning: A Program to Improve Children's Reading Awareness and Comprehension." *Journal of Educational Psychology* 76 (1984): 1239–52.

Park, Hyekyung, and Shinobu Kitayama. "Perceiving through Culture: The Socialized Attention Hypothesis." In Reginald B. Adams Jr. et al. (eds.), *The Science of Social Vision*, 75–89. New York: Oxford University Press, 2011.

Parker, Ashley, and Michael Barbaro. "Romney Takes Analytic Approach to Campaign Chaos." *New York Times*, February 28, 2012, A16.

Pashler, Harold E. *The Psychology of Attention*. Cambridge, MA: MIT Press, 1999.

Pease, Barbara, and Allan Pease. *Why Men Don't Listen and Women Can't Read Maps: How We're Different and What to Do about It*. New York: Broadway Books, 2001 [2000].

Pesce-Anzeneder, Caterina P., and Rainer Bösel. "Modulation of the Spatial Extent of the Attentional Focus in High-Level Volleyball Players." *European Journal of Cognitive Psychology* 10 (1998): 247–67.

Peterson, Mary A., and Bradley S. Gibson. "The Initial Identification of Figure-Ground Relationships: Contributions from Shape Recognition Processes." *Bulletin of the Psychonomic Society* 29 (1991): 199–202.

Pilch, John J. *Flights of the Soul: Visions, Heavenly Journeys, and Peak Experiences in the Biblical World*. Grand Rapids, MI: William B. Eerdmans, 2011.

Pind, Jörgen L. "Figure and Ground at 100." *The Psychologist* 25, no. 1 (January 2012): 90–91. http://www.psy.ku.dk/om/Historie/figure_and_ground_at_100/JLPind-Psychologist.pdf/.

Pirandello, Luigi. *Tonight We Improvise*. New York: Samuel French, 1960 [1932].

Plaisted, Kate, et al. "Enhanced Visual Search for a Conjunctive Target in Autism: A Research Note." *Journal of Child Psychology and Psychiatry* 39 (1998): 777–83.

Pliny the Elder. *The Natural History of Pliny*. London: Henry G. Bohn, 1857 [circa AD 77].

Pluta, Stefan A. "Affidavit in Support of Criminal Complaint, Arrest Warrant and Search Warrants against Robert Philip Hanssen, February 2001." http://www.fbi.gov/about-us/history/famous-cases/robert-hanssen/affidavit.pdf.

Poe, Edgar A. "The Purloined Letter." In *A Collection of Stories*, 188–208. New York: Tom Doherty Associates, 1988 [1844].

Portmann, Adolf. *Animal Camouflage*. Ann Arbor: University of Michigan Press, 1959.

Posner, Michael. "Psychobiology of Attention." In Michael S. Gazzaniga and Colin Blakemore (eds.), *Handbook of Psychobiology*, 441–80. New York: Academic Press, 1975.

Posner, Michael I., et al. "Attention and the Detection of Signals." *Journal of Experimental Psychology: General* 109 (1980): 160–74.

Poulton, Edward B. *The Colours of Animals*. New York: D. Appleton, 1890.

Proctor, Robert N. "Agnotology: A Missing Term to Describe the Cultural Production of Ignorance (and Its Study)." In Robert N. Proctor and Londa Schiebinger (eds.), *Agnotology: The Making and Unmaking of Ignorance*, 1–33. Stanford, CA: Stanford University Press, 2008.

Purcell, Carl. *Your Artist's Brain*. Cincinnati, OH: North Light Books, 2010.

Purcell, Kristen. "In a League of Their Own: Mental Leveling and the Creation of Social Comparability in Sport." *Sociological Forum* 11 (1996): 421–33.

Pylyshyn, Zenon W., and Ron W. Storm. "Tracking Multiple Independent Targets: Evidence for a Parallel Tracking Mechanism." *Spatial Vision* 3 (1988): 179–97.

Reingold, Eyal M., and Neil Charness. "Perception in Chess: Evidence from Eye Movements." In Geoffrey Underwood (ed.), *Cognitive Processes in Eye Guidance*, 325–54. New York: Oxford University Press, 2005.

Reit, Seymour. *Masquerade: The Amazing Camouflage Deceptions of World War II*. New York: Hawthorn Books, 1978.

Ren, Xiaofeng, et al. "Figure/Ground Assignment in Natural Images." *Lecture Notes in Computer Science* 3952 (2006): 614–27.

Rensink, Ronald A., et al. "To See or Not to See: The Need for Attention to Perceive Changes in Scenes." *Psychological Science* 8 (1997): 368–73.

Roehler, Laura R., and Gerald G. Duffy. "Direct Explanation of Comprehension Processes." In Gerald G. Duffy (ed.), *Comprehension Instruction: Perspectives and Suggestions*, 265–80. New York: Longman, 1984.

Rogoff, Barbara. *Apprenticeship in Thinking: Cognitive Development in Social Context.* New York: Oxford University Press, 1990.

Rogoff, Barbara, et al. "Toddlers' Guided Participation with Their Caregivers in Cultural Activity." In Ellice A. Forman et al. (eds.), *Contexts for Learning: Sociocultural Dynamics in Children's Development,* 230–53. New York: Oxford University Press, 1996.

Rosenbloom, Joseph. *Doctor Knock-Knock's Official Knock-Knock Dictionary.* New York: Sterling, 1976.

Rubin, Edgar. "Figure and Ground." In David C. Beardslee and Michael Wertheimer (eds.), *Readings in Perception,* 194–203. Princeton, NJ: D. Van Nostrand Company, 1958 [1915].

———. *Visuell Wahrgenommene Figuren: Studien in Psychologischer Analyse.* Copenhagen: Gyldendalske Boghandel, 1921 [1915].

Rubin, Jeffrey B. "Deepening Listening: The Marriage of Buddha and Freud." In Uwe P. Gielen et al. (eds.), *Principles of Multicultural Counseling and Therapy,* 373–89. New York: Routledge, 2008.

Rubin, Nava. "Figure and Ground in the Brain." *Nature Neuroscience* 4 (2001): 857–58.

Ruxton, Graeme D. "Non-Visual Crypsis: A Review of the Empirical Evidence of Camouflage to Senses Other than Vision." *Philosophical Transactions of the Royal Society B: Biological Sciences* 364 (2009): 549–57.

Saenger, Paul. *Space between Words: The Origins of Silent Reading.* Stanford, CA: Stanford University Press, 1997.

Saint-Amour, Paul K. "Modernist Reconnaissance." *Modernism/Modernity* 10 (2003): 349–80.

Salvucci, Dario D., and Niels A. Taatgen. *The Multitasking Mind.* New York: Oxford University Press, 2011.

Schachtel, Ernest G. *Metamorphosis: On the Development of Affect, Perception, Attention, and Memory.* New York: Basic Books, 1959.

Schafer, Roy, and Gardner Murphy. "The Role of Autism in a Visual Figure-Ground Relationship." *Journal of Experimental Psychiatry* 32 (1943): 335–43.

Schattschneider, Doris. *Visions of Symmetry: Notebooks, Periodic Drawings, and Related Work of M. C. Escher.* New York: W. H. Freeman, 1990.

Schechner, Richard. *Environmental Theater.* New York: Hawthorn, 1973.

Schiffman, Nathaniel. *Abracadabra: Secret Methods Magicians and Others Use to Deceive Their Audience.* Amherst, NY: Prometheus Books, 1997.

Schlesinger, Herbert J. "Cognitive Attitudes in Relation to Susceptibility to Interference." *Journal of Personality* 22 (1954): 354–74.

Schocken, Deganit S. *How Many Is One.* Tel Aviv, Israel: Tel Aviv Museum of Art, 2003.

Schuette, Sarah L. *Pets All Around: A Spot-It Challenge.* North Mankato, MN: Capstone Press, 2013.

———. *Sports Zone: A Spot-It Challenge.* North Mankato, MN: Capstone Press, 2013.

Schutz, Alfred. "Making Music Together: A Study in Social Relationship," *Social Research* 18 (1951): 76–97.

Schutz, Alfred, and Thomas Luckmann. *The Structures of the Life-World.* Evanston, IL: Northwestern University Press, 1973.

Schwitzgebel, Robert. "The Performance of Dutch and Zulu Adults in Selected Perceptual Tasks." *Journal of Social Psychology* 57 (1962): 73–77.

Seelye, Katharine Q., and Ian Lovett. "After Attack, Suspects Returned to Routines, Raising No Suspicions." *New York Times,* April 27, 2013, A12.

Seelye, Katharine Q., et al. "F.B.I. Posts Images of Pair Suspected in Boston Attack." *New York Times,* April 19, 2013, A1, A20. http://www.nytimes.com/2013/04/19/us/fbi-releases-video-of-boston-bombing-suspects.html?_r=0&pagewanted=print (accessed on August 7, 2013).

Segil, Larraine. *Dynamic Leader, Adaptive Organization: Ten Essential Traits for Managers.* New York: John Wiley & Sons, 2002.

Shah, Amitta, and Uta Frith. "An Islet of Ability in Autistic Children: A Research Note." *Journal of Child Psychology and Psychiatry and Allied Disciplines* 24 (1983): 613–20.

Shell, Hanna R. *Hide and Seek: Camouflage, Photography, and the Media of Reconnaissance.* New York: Zone Books, 2012.

Shepard, Roger N. *Mind Sights: Original Visual Illusions, Ambiguities, and Other Anomalies, with a Commentary on the Play of Mind in Perception and Art.* New York: W. H. Freeman and Company, 1990.

Sherman, Julia A. "Problem of Sex Differences in Space Perception and Aspects of Intellectual Functioning." *Psychological Review* 74 (1967): 290–99.

Shklovsky, Victor. "Art as Technique." In Lee T. Lemon and Marion J. Reis (eds.), *Russian Formalist Criticism: Four Essays*, 5–24. Lincoln: University of Nebraska Press, 1965 [1917].

Shlain, Leonard. *The Alphabet versus the Goddess: The Conflict between Word and Image.* New York: Penguin Arkana, 1998.

Shneidman, Edwin S. *The Suicidal Mind.* New York: Oxford University Press, 1996.

Silver, Ruth. *Invisible: My Journey through Vision and Hearing Loss.* iUniverse, 2012.

Silverman, Julian. "Scanning-Control Mechanism and 'Cognitive Filtering' in Paranoid and Non-Paranoid Schizophrenia." *Journal of Consulting Psychology* 28 (1964): 385–93.

Simmel, Georg. "The Field of Sociology." In Kurt H. Wolff (ed.), *The Sociology of Georg Simmel*, 3–25. New York: Free Press, 1950 [1917].

Simons, Daniel J., and Christopher F. Chabris. "Gorillas in Our Midst: Sustained Inattentional Blindness for Dynamic Events." *Perception* 28 (1999): 1059–74.

Singer, Peter. *The Expanding Circle: Ethics and Sociobiology.* New York: Farrar, Straus & Giroux, 1981.

Sloboda, John, and Judy Edworthy. "Attending to Two Melodies at Once: The Effect of Key Relatedness." *Psychology of Music* 9 (1981): 39–43.

Smith, Barbara H. *Poetic Closure: A Study of How Poems End.* Chicago: University of Chicago Press, 1968.

Sofsky, Wolfgang. *The Order of Terror: The Concentration Camp.* Princeton, NJ: Princeton University Press, 1996 [1993].

Sorensen, Roy. *Seeing Dark Things: The Philosophy of Shadows.* New York: Oxford University Press, 2008.

Sowden, Paul T., et al. "Perceptual Learning of the Detection of Features in X-Ray Images: A Functional Role for Improvements in Adults' Visual Sensitivity?" *Journal of Experimental Psychology: Human Perception and Performance* 26 (2000): 379–90.

Spelke, Elizabeth S., et al. "The Development of Object Perception." In Stephen M. Kosslyn and Daniel N. Osherson (eds.), *Visual Cognition: An Invitation to Cognitive Science.* Vol. 2, 297–330. 2nd ed. Cambridge, MA: MIT Press, 1995.

Spender, Dale. *Man Made Language.* London: Routledge & Kegan Paul, 1980.

Spitzer, Gabriel. "Clever Apes #29: Nature and Human Nature." http://www.wbez.org/blogs/clever-apes/2012-04/clever-apes-29-nature-and-human-nature-97867 (accessed on January 2, 2013).

Srinivasan, Narayanan, et al. "Focused and Distributed Attention." *Progress in Brain Research* 176 (2009): 87–100.

Stalnaker, Robert. "Presuppositions." *Journal of Philosophical Logic* 2 (1973): 447–57.

Steinkraus, Warren E. "The Art of Conjuring." *Journal of Aesthetic Education* 13, no. 4 (October 1979): 17–27.

Stevens, Martin, and Sami Merilaita. "Animal Camouflage: Current Issues and New Perspectives." *Philosophical Transactions of the Royal Society B: Biological Sciences* 364 (2009): 423–27.

———. "Defining Disruptive Coloration and Distinguishing Its Functions." *Philosophical Transactions of the Royal Society B: Biological Sciences* 364 (2009): 481–88.

Stockwell, Peter. *Cognitive Poetics: An Introduction*. London: Routledge, 2002.

Stone, Christopher. *Should Trees Have Standing? Toward Legal Rights for Natural Objects*. Los Altos, CA: William Kaufmann, 1974.

Stoppard, Tom. *Rosencrantz and Guildenstern Are Dead*. New York: Samuel French, 1995 [1967].

Subirana-Vilanova, J. Brian, and Whitman Richards. "Attentional Frames, Frame Curves and Figural Boundaries: The Inside/Outside Dilemma." *Vision Research* 36 (1996): 1493–1501.

Sullivan, Harry S. *Clinical Studies in Psychiatry*. New York: W. W. Norton, 1956.

Sykes, Gresham M., and David Matza. "Techniques of Neutralization: A Theory of Delinquency." *American Sociological Review* 22 (1957): 664–70.

Sypher, Wylie. *Rococo to Cubism in Art and Literature*. New York: Random House, 1960.

Tagg, Philip. "Accompaniment." In John Shepherd et al. (eds.), *Continuum Encyclopedia of Popular Music of the World*. Vol. II, *Performance and Production*, 628–30. London: Continuum, 2003.

———. "Melody and Accompaniment" (no date). http://www.andrelambert.org/uqam/principes/melodaccUS.pdf (accessed on August 1, 2012).

Taussig, Michael. *Defacement: Public Secrecy and the Labor of the Negative*. Stanford: Stanford University Press, 1999.

Teale, William H., and Miriam G. Martinez. "Reading Aloud to Young Children: Teachers' Reading Styles and Kindergartners' Text Comprehension." In Clotilde Pontecorvo et al. (eds.), *Children's Early Text Construction*, 321–44. Mahwah, NJ: Lawrence Erlbaum, 1996.

Teshigahara, Wafu. *Ikebana: The New Illustrated Guide to Mastery*. Tokyo: Kodansha International, 1966.

Thayer, Abbott H. "Disruptive Camouflage." In Roy R. Behrens (ed.), *Ship Shape: A Dazzle Camouflage Sourcebook*, 35–43. Dysart, IA: Bobolink Books, 2012 [1918].

Thayer, Gerald, and Abbott H. Thayer. *Concealing-Coloration in the Animal Kingdom: An Exposition of the Laws of Disguise through Color and Pattern*. New York: Macmillan, 1909.

Titchener, Edward B. *A Text-Book of Psychology*. New York: Macmillan, 1910.

Tobin, Jacqueline L., and Raymond G. Dobard. *Hidden in Plain View: A Secret Story of Quilts and the Underground Railroad*. New York: Anchor Books, 2000 [1999].

Townsend, James T. "Serial vs. Parallel Processing: Sometimes They Look like Tweedledum and Tweedledee but They Can (and Should) Be Distinguished." *Psychological Science* 1 (1990): 46–54.

Townsend, Jeanne, and Eric Courchesne. "Parietal Damage and Narrow 'Spotlight' Spatial Attention." *Journal of Cognitive Neuroscience* 6 (1994): 220–32.

Trancik, Roger. *Finding Lost Space: Theories of Urban Design*. New York: John Wiley & Sons, 1986.

Treisman, Anne M. "How the Deployment of Attention Determines What We See." *Visual Cognition* 14 (2006): 411–43.

———. "Strategies and Models of Selective Attention." *Psychological Review* 76 (1969): 282–99.

Troscianko, Tom, et al. "Camouflage and Visual Perception." *Philosophical Transactions of the Royal Society B: Biological Sciences* 364 (2009): 449–61.

Trubetzkoy, Nikolai S. *Principles of Phonology*. Berkeley: University of California Press, 1969 [1939].

Truffaut, François. *Hitchcock*. Rev. ed. New York: Touchstone Books, 1985 [1983].

Tuan, Yi-Fu. *Segmented Worlds and Self*. Minneapolis: University of Minnesota Press, 1982.

Tzu, Sun. *The Art of War.* Oxford: Oxford University Press, 1963 [sixth century BC].

Urry, John. *The Tourist Gaze: Leisure and Travel in Contemporary Societies.* London: SAGE Publications, 1990.

Uskul, Ayse K., et al. "Ecocultural Basis of Cognition: Farmers and Fishermen Are More Holistic Than Herders." *Proceedings of the National Academy of Sciences of the United States of America* 105 (June 24, 2008): 8552–56.

Uspensky, Boris. *A Poetics of Composition.* Berkeley: University of California Press, 1973.

Van Peer, Willie. *Stylistics and Psychology: Investigations of Foregrounding.* London: Croom Helm, 1986.

Varela, Francisco J., and Natalie Depraz. "Wisdom Traditions and the Ways of Reduction." In Natalie Depraz et al. (eds.), *On Becoming Aware: A Pragmatics of Experiencing*, 205–31. Amsterdam, The Netherlands: John Benjamins Publishing Company, 2003.

Vaznis, Bill. *500 Deer Hunting Tips: Strategies, Techniques, and Methods.* Minneapolis, MN: Creative Publishing International, 2008.

Vernon, Magdalen D. *The Psychology of Perception.* Harmondsworth, UK: Penguin, 1962.

———. *Visual Perception.* Ann Arbor: University of Michigan Press, 1937.

Wachtel, Paul L. "Conceptions of Broad and Narrow Attention." *Psychological Bulletin* 68 (1967): 417–29.

Wackermann, Jirí, et al. "Ganzfeld-Induced Hallucinatory Experience, Its Phenomenology and Electrophysiology." *Cortex* 44 (2008): 1364–78.

Wallace, Alfred R. *Darwinism: An Exposition of the Theory of Natural Selection with Some of Its Applications.* London: Macmillan, 1891.

Wårvik, Brita. "What Is Foregrounded in Narratives? Hypotheses for the Cognitive Basis of Foregrounding." In Tuija Virtanen (ed.), *Approaches to Cognition Through Text and Discourse*, 99–122. Berlin: Mouton de Gruyter, 2004.

Watson, Colin. *Lonelyheart 4122: A Flaxborough Novel.* Bath, UK: Chivers Press, 1993.

Watson, Ian. "How Tour Operators and Travel Guidebooks Select Destinations." Lecture given at the "Practicing Nature-Based Tourism" Conference, Reykjavik Art Museum, February 2011.

Watts, Alan. *The Two Hands of God: The Myths of Polarity.* Collier Books, 1969.
Waugh, Linda. "Marked and Unmarked: A Choice between Unequals in Semiotic Structure." *Semiotica* 38 (1982): 299–318.
Wells, Herbert G. *Italy, France and Britain at War.* New York: Macmillan, 1917.
Werner, Heinz, and Seymour Wapner. "Toward a General Theory of Perception." In David C. Beardslee and Michael Wertheimer (eds.), *Readings in Perception*, 491–512. Princeton, NJ: D. Van Nostrand Company, 1958 [1952].
Wever, Ernest G. "Attention and Clearness in the Perception of Figure and Ground." *American Journal of Psychology* 40 (1928): 51–74.
Williams, A. Mark. "Perceptual Skill in Soccer: Implications for Talent Identification and Development." *Journal of Sports Sciences* 18 (2000): 737–50.
Williams, A. Mark, et al. "Visual Search Strategies in Experienced and Inexperienced Soccer Players." *Research Quarterly for Exercise and Sport* 65 (1994): 127–35.
Williams, Henry S. "Natural Selection and Ship Camouflage." In Roy R. Behrens (ed.), *Ship Shape: A Dazzle Camouflage Sourcebook*, 81–95. Dysart, IA: Bobolink Books, 2012 [1919].
Witkin, Herman A., and John W. Berry. "Psychological Differentiation in Cross-Cultural Perspective." *Journal of Cross-Cultural Psychology* 6 (1975): 4–87.
Witkin, Herman A., et al. *Personality through Perception: An Experimental and Clinical Study.* New York: Harper & Brothers, 1954.
———. *Psychological Differentiation: Studies of Development.* New York: John Wiley & Sons, 1962.
Wittgenstein, Ludwig. *Philosophical Investigations.* 4th ed. Malden, MA: Wiley-Blackwell, 2009 [1953].
Wolfe, Jeremy M., et al. "Segmentation of Objects from Backgrounds in Visual Search Tasks." *Vision Research* 42 (2002): 2985–3004.
Wong, Eva, and Naomi Weisstein. "Sharp Targets Are Detected Better against a Figure, and Blurred Targets Are Detected Better against a Background." *Journal of Experimental Psychology: Human Perception and Performance* 9 (1983): 194–202.
Wood, Beverly P. "Visual Expertise." *Radiology* 211 (1999): 1–3.
Wood, Lindsay. "How to Remove Scent from Hunting Gear." http://www.ehow.com/how_5663627_remove-scent-hunting-gear.html.

Zakia, Richard D. *Perception and Imaging.* 2nd ed. Boston: Focal Press, 2002.

Zangemeister, W. H., et al. "Evidence for a Global Scanpath Strategy in Viewing Abstract Compared with Realistic Images." *Neuropsychologia* 33 (1995): 1009–25.

Zerubavel, Eviatar. *Ancestors and Relatives: Genealogy, Identity, and Community.* New York: Oxford University Press, 2011.

———. *The Elephant in the Room: Silence and Denial in Everyday Life.* New York: Oxford University Press, 2006.

———. *The Fine Line: Making Distinctions in Everyday Life.* Chicago: University of Chicago Press, 1993 [1991].

———. *Hidden Rhythms: Schedules and Calendars in Social Life.* Berkeley: University of California Press, 1985 [1981].

———. "Horizons: On the Sociomental Foundations of Relevance." *Social Research* 60 (1993): 397–413.

———. "Language and Memory: 'Pre-Columbian' America and the Social Logic of Periodization." *Social Research* 65 (1998): 315–30.

———. *Patterns of Time in Hospital Life: A Sociological Perspective.* Chicago: University of Chicago Press, 1979.

———. "Personal Information and Social Life." *Symbolic Interaction* 5, no.1 (1982): 97–109.

———. *Social Mindscapes: An Invitation to Cognitive Sociology.* Cambridge, MA: Harvard University Press, 1997.

———. *The Seven-Day Circle: The History and Meaning of the Week.* Chicago: University of Chicago Press, 1989 [1985].

———. "The Social Marking of the Past: Toward a Socio-Semiotics of Memory." In Roger Friedland and John Mohr (eds.), *Matters of Culture: Cultural Sociology in Practice,* 184–95. Cambridge: Cambridge University Press, 2004.

———. *Time Maps: Collective Memory and the Social Shape of the Past.* Chicago: University of Chicago Press, 2003.

Zerubavel, Noga. "Restricted Awareness in Intimate Partner Violence: The Effect of Childhood Sexual Abuse and Fear of Abandonment." Ph.D. diss., Miami University, 2013.

Zerubavel, Yael. "Space Metaphors in Modern Israeli Culture." Paper presented at the Annual Meeting of the Association for Jewish Studies, Chicago, 2012.

AUTHOR INDEX

Ackerman, Diane, 111
Alario, Margarita, 120, 121
Alvarez, George A., 136
Alves, Stephanie E., 123
Amis, Martin, 129
Arnheim, Rudolf, 101, 103, 104, 105, 108, 141, 147
Arvidson, P. Sven, 96, 138
Aulls, Mark W., 130, 141
Austin, James H., 96, 97, 139, 140, 148
Avant, Lloyd L., 111

Bacigalupe, Gonzalo, 100
Baden-Powell, Robert, 30, 113
Bakhtiarov, Oleg, 143
Bankier, David, 129
Barbaro, Michael, 107
Baron-Cohen, Simon, 143
Bartlett, Frederic C., 123
Bateson, Gregory, 21, 108
Bateson, Mary C., 122
Baumann, James F., 130, 141
Baumgartner, Frank R., 133

Baylis, Gordon C., 103, 104, 146
Bazerman, Max H., 95
Beck, John C., 99, 110, 133
Becklen, Robert, 98
Beddard, Frank E., 33, 115
Behrens, Roy R., 114, 115, 116, 117, 118, 119, 120
Berger, Harris M., 107, 138
Berger, Peter L., 126, 133
Bergson, Henri, 96
Berlyne, D. E., 135
Berry, John W., 118, 119, 121, 122, 123, 124
Bichot, Narcisse P., 98
Bieri, James, 122
Bigand, E., 110, 137
Binet, Alfred, 4, 98
Birrell, Pamela, 100
Blake, William, 74, 136
Blechman, Hardy, 115, 116, 117, 118
Bleuler, Eugen, 135
Boduroglu, Aysecan, 125, 126
Bogdashina, Olga, 112, 135, 142, 143
Bosk, Charles L., 133

Bösel, Rainer, 139
Bothwell, Dorr, 147
Bourdieu, Pierre, 130
Brandt, Edwin H., 117
Bregman, Albert S., 18, 105, 106, 108, 110, 111, 131, 137
Breitmeyer, Bruno G., 121
Brekhus, Wayne H., 108, 109, 111, 112, 113, 144, 145
Bressan, Paola, 99
Broadbent, Donald E., 99, 105
Brochard, Renaud, 106, 131
Brown, Thomas E., 142
Brown, Tom, 1, 95, 138, 144, 145, 148
Brunius, Teddy, 147
Burch, Bridgette, 120
Burkan, Wayne C., 99
Buswell, Guy T., 126
Butor, Michel, 147

Cahill, Spencer E., 128
Cale, Ellison M., 104
Campbell, Jeffrey, 106
Campion, Lisa, 127
Caplan, Paula J., 135, 142
Carson, Shelley H., 140
Casati, Roberto, 102, 105, 109
Cassidy, Steve, 106
Casson, Lionel, 118
Cavanagh, Patrick, 136
Cave, Kyle R., 98
Cerulo, Karen A., 99
Cervone, Daniel, 98
Chabris, Christopher F., 4, 7, 46, 96, 97, 98, 99, 100
Chambers, Ross, 108, 111
Chapman, James, 135, 141
Chapman, Steve, 116
Charness, Neil, 131
Chavajay, Pablo, 124

Cherry, E. Colin, 106
Chevan, David, 146
Childers, Doug, 96
Chong, Sang C., 137
Chua, Hannah F., 125, 126
Chugh, Dolly, 95
Cobb, Roger W., 133
Cohen, Bernard C., 71, 134
Cohen, Stanley, 129
Cohen, Walter, 110
Collins, Allan, 130
Collins, Michael, 116
Copland, Aaron, 20, 138
Correa-Chávez, Maricela, 124
Cott, Hugh B., 147
Courchesne, Eric, 97, 143
Crary, Jonathan, 96, 97, 100, 126, 133, 142
Crundall, David, 132, 139
Cummings, Edward E., 91, 148
Cunningham, James W., 128, 141
Cuthill, Innes, C., 114, 117

Darwin, Erasmus, 32, 33, 114
Davenport, Thomas H., 99, 110, 133
Davey, Beth, 130
Davis, Murray S., 123
Day, George S., 99, 139, 149
Deikman, Arthur J., 144, 148
Dellas, Marie, 140
Dennett, Daniel C., 114
Denton, Jeremiah A., 37, 117
Depraz, Natalie, 138, 139, 140
Derber, Charles, 110
DesAutels, Peggy, 99, 107, 144
DeVault, Marjorie L., 129
Dewey, John, 2, 3, 96
Dilthey, Wilhelm, 3, 97
Divenyi, Pierre L., 105, 110, 119
Dobard, Raymond G., 116
Doherty, Martin J., 122, 126, 136

Downs, Anthony, 134
Doyle, Arthur C., 84, 145
Driver, Jon, 104, 146
Duffy, Gerald G., 130
Duncan, Andrew, 112
Duncan, John, 113, 118
Dunn, Bruce R., 97, 140
Duranti, Alessandro, 102, 107
Durkheim, Emile, 4, 98, 127, 128
Dykes, Margaret, 140, 141

Edwards, Betty, 85, 86, 145
Edworthy, Judy, 131, 138
Egeth, Howard E., 111
Ehrenzweig, Anton, 76, 137, 145, 146, 147, 148
Ekman, Paul, 144
Elder, Charles D., 133
Ellison, Ralph, 112
Emerson, Joan P., 128
Endler, John A., 114
Escher, Maurits C., 86, 89, 103, 106, 146
Evans, Karla K., 137

Fehmi, Les, 136, 145, 147, 149
Felman, Shoshana, 129
Féré, Charles, 98
Fernald, Anne, 130
Fernandez-Duque, Diego, 98, 100
Findlay, John M., 120
Fine, Gary A., 110
Fingarette, Herbert, 129, 145
Fisher, Charles, 148
Fleck, Ludwik, 132
Fleischman, Suzanne, 142, 147
Follette, Victoria M., 140
Force, William R., 113
Foucault, Michel, 128
Francolini, Carl M., 111

Freeman, Thomas, 135
Freud, Anna, 6, 100
Freud, Sigmund, 58, 127
Freudenburg, William R., 120, 121
Freyd, Jennifer J., 100
Friedman, Asia, 6, 49, 99, 100, 109, 121, 123, 126, 127, 130
Friedman, Ronald S., 99, 107, 140
Friend, Trudy, 102
Friesen, Wallace V., 144
Frith, Uta, 143
Furley, Philip, 98, 119, 135, 136, 139, 140

Gaier, Eugene L., 140
Gans, Herbert J., 134
Garfinkel, Harold, 84, 111, 145
Garland-Thomson, Rosemarie, 120, 128, 142
Ghent, Lila, 118
Gibson, Bradley S., 103, 104
Gibson, James J., 101, 102, 121, 141, 147
Giedion, Sigfried, 137
Gilligan, Carol, 107
Goffman, Erving, 28, 61, 69, 84, 108, 112, 123, 127, 128, 133, 145
Gold, Steven N., 100
Gombrich, Ernst H., 102
Goodwin, Charles, 102, 107, 110, 126
Gottschaldt, Kurt, 43, 119
Gottsdanker, Robert, 113
Grandin, Temple, 82, 121, 143
Grauds, Constance, 96
Grazian, David, 129
Graziano, Frank, 129
Green, John, 102
Gregory, Andrew H., 110
Griffiths, Tom, 99
Grossman, Valerie G. A., 132

Gugerty, Leo, 137
Gurwitsch, Aron, 104, 138
Gutchess, Angela H., 125, 126

Haaken, Janice, 124, 136
Hall, Edward T., 58, 124, 127
Halley, Richard D., 122
Halton, Eugene, 138
Handford, Martin, 119
Happé, Francesca, 142, 143
Harrington, Jonathan, 106
Hartmann, Thom, 139
Hawkins, Stan, 107
Hayes, Steven C., 136
Heaton, Pamela F., 143
Heilman, Samuel C., 123
Heller, Joseph, 91, 148
Helson, Harry, 101, 102, 103, 147
Hernández-Peón, Raúl, 98, 99, 107
Hilgartner, Stephen, 133
Hirschauer, Stefan, 126
Hirsh, Ira J., 105, 110, 119
Hochberg, Julian E., 103, 105, 109
Hochschild, Arlie, 131
Hodges, Donald A., 106
Hofstadter, Douglas R., 103, 106, 111, 146
Hopper, Paul J., 108
Horn, David, 107
Horowitz, Alexandra, 84, 96, 134, 135, 144
Horowitz, Seth S., 144
Horwitz, Allan V., 108
Horwitz, Gordon J., 129
Hotchkiss, John, 113
Howell, Elizabeth F., 100
Hsiao, Steven S., 98, 99, 141
Hubert, Henri, 108
Huff, Richard, 117
Humphreys, Glyn W., 113, 118
Huron, David, 110

Husserl, Edmund, 85, 145
Huxley, Aldous, 2, 96

Iserson, Kenneth V., 132
Iwasaki, Syoichi, 138
Iyengar, Shanto, 134

Jaarsma, Pier, 142
James, William, 2, 11, 93, 96, 98, 101, 108, 112, 149
Jastrow, Joseph, 4, 11, 95, 97, 99, 101, 110, 138
Jessel, David, 122
Ji, Li-Jun, 125
Johnson, Catherine, 121, 143
Johnson, Mark, 98, 100, 133
Jolliffe, Therese, 143
Jones, Bryan D., 133
Jones, Dave, 139
Jones, Warren, 122, 142
Joseph, Eddie, 120
Julesz, Bela, 138
Jung, Richard, 97, 110

Kabat-Zinn, Jon, 140, 144
Kahneman, Daniel, 101, 104, 106, 123
Kanizsa, Gaetano, 102, 103, 104, 141
Kaprow, Allan, 148
Kashatus, William C., 116
Kasof, Joseph, 121, 140, 141
Kates, Brian, 117
Katz, David, 105
Keefe, Thomas, 96
Keen, Angeline M., 114, 115
Keller, Heidi, 133
Kelner, Shaul, 129
Kendon, Adam, 133

Kepes, Gyorgy, 102, 146, 147
Kern, Stephen, 102, 115, 116, 119, 137, 146, 148
Kinder, Donald R., 134
Kingdon, John W., 134
Kinsbourne, Marcel, 135, 142
Kirby, Michael, 148
Kirby, Victoria N., 148
Kitayama, Shinobu, 124, 125, 126, 129
Kjolseth, Rolf, 111
Klin, Ami, 122, 142
Klinenberg, Eric, 134
Koffka, Kurt, 102, 103, 104, 105, 106, 109, 118, 132, 138, 141, 147
Köhler, Wolfgang, 102, 103, 105, 114, 118, 119, 141
Kohn, Livia, 96, 139
Kostelanetz, Richard, 148
Kostelnick, Charles, 141
Krupinski, Elizabeth A., 118
Kubovy, Michael, 103
Kuhn, Gustav, 120
Kuhn, Thomas S., 58, 127, 132, 143
Kühnen, Ulrich, 124
Kuiken, Don, 144
Kundel, Harold L., 109, 118
Kuran, Timur, 134
Kusakov, Igor, 139, 143

LaBerge, David, 97
Lakoff, George, 133
Lamont, Peter, 120
Lane, David M., 130, 141
Lang, John T., 127
Lantigua-Williams, Juleyka, 134
Lanza, Joseph, 107
Lanzmann, Claude, 129
Laqueur, Thomas, 58, 127
Laub, Dori, 129
Leder, Drew, 98, 99, 138

Lesgold, Alan, 132
Lewis, Richard S., 125
Lind, Johan, 117
Linden, William, 119
Lovett, Ian, 112
Luchterhand, Elmer, 128, 129
Luckiesh, Matthew, 118
Luckmann, Thomas, 111, 126, 133

Maccoby, Eleanor, 121
MacDonald, Norma, 135, 141
Mack, Arien, 96, 98, 99, 132
Macknik, Stephen L., 120
Mac Low, Jackson, 89, 146
Mallarmé, Stéphane, 89, 146
Markus, Hazel R., 124
Martin, Judith, 128
Martinez, Miriam G., 130, 142
Mason-Schrock, Douglas, 128
Masuda, Takahiko, 125, 126
Matza, David, 100
Mayfield, Marlys, 147
Mazza, Veronica, 103
McAdams, Stephen, 106
McCombs, Maxwell E., 133, 134
McGhie, Andrew, 135, 140, 141
Mello-Thoms, Claudia, 132
Memmert, Daniel, 98, 119, 131, 135, 136, 139, 140
Merilaita, Sami, 113, 117
Merleau-Ponty, Maurice, 101, 102, 105
Metzger, Wolfgang, 95, 101, 102, 103, 105, 114, 117, 132, 146
Meyrowitz, Joshua, 134
Miall, David S., 144
Miller, Leon K., 143
Mills, C. Wright, 132
Mills, Charles W., 6, 100
Milner, Marion, 72, 74, 134, 136, 148, 149

Miner, Horace, 84, 145
Minissale, Gregory, 131
Mishler, Elliot G., 136
Miyamoto, Yuri, 125
Moir, Anne, 122
Monahan, Sean, 116
Moore, Cathleen, 99
Moore, David W., 128, 141
Moran, Aidan, 136
Morikawa, Hiromi, 130
Moskop, John C., 132
Mourant, Ronald R., 132
Murphy, Gardner, 140

Neisser, Ulric, 24, 97, 98, 106, 110
Newark, Tim, 115, 116, 117
Nisbett, Richard E., 55, 99, 122, 125
Nodine, Calvin F., 109, 118, 119, 131, 132
Norgaard, Kari M., 123
Nougier, Vincent, 139

Ocasio, William, 133
Ochs, Elinor, 124
Odets, Clifford, 91, 148
Olsen, Mark, 83, 144
Ornstein, Robert, 96, 110, 111, 144

Panek, Richard, 143
Paris, Scott, 130, 141
Park, Hyekyung, 126, 129
Parker, Ashley, 107
Pashler, Harold E., 97
Pearson, Deborah A., 130, 141
Pease, Allan, 122
Pease, Barbara, 122
Pesce-Anzeneder, Caterina P., 139
Peterson, Mary A., 103, 104
Pilch, John J., 97

Pind, Jörgen L., 101, 104, 109
Pirandello, Luigi, 91, 148
Pizzighello, Silvia, 99
Plaisted, Kate, 118
Pliny the Elder, 54, 124
Pluta, Stefan A., 17
Poe, Edgar A., 27, 111
Portmann, Adolf, 114
Posner, Michael I., 97
Poulton, Edward B., 32, 33, 114, 115
Proctor, Robert N., 100
Purcell, Carl, 24, 90, 109, 111, 113, 146, 147
Purcell, Kristen, 126
Pylyshyn, Zenon W., 137

Reingold, Eyal M., 131
Reit, Seymour, 113, 117, 120, 121
Ren, Xiaofeng, 104
Rensink, Ronald A., 98, 141
Richards, Whitman, 101, 105
Robbins, Jim, 136, 145, 147, 149
Roberts, David D., 141
Rock, Irvin, 96, 98, 99
Rockwell, Thomas H., 132
Roehler, Laura R., 130
Rogoff, Barbara, 113, 124
Rosenbloom, Joseph, 119
Rubin, Edgar, 11, 14, 16, 101, 102, 103, 104, 105, 109
Rubin, Jeffrey B., 97, 139
Rubin, Nava, 104
Ruxton, Graeme D., 113, 116

Saenger, Paul, 106
Saint-Amour, Paul K., 115, 117, 119, 132, 137
Salvucci, Dario D., 137
Schachtel, Ernest G., 97
Schafer, Roy, 140

Schattschneider, Doris, 146
Schechner, Richard, 148
Schiffman, Nathaniel, 120
Schlesinger, Herbert J., 96, 121, 135, 141
Schocken, Deganit S., 147
Schoemaker, Paul J. H., 99, 139, 149
Schuette, Sarah L., 119
Schutz, Alfred, 111, 133
Schwitzgebel, Robert, 124
Sebald, David C., 106
Seelye, Katharine Q., 111, 112
Segil, Larraine, 145
Shah, Amitta, 143
Shaw, Donald L., 134
Shell, Hanna R., 114, 115
Shepard, Roger N., 86, 87, 146
Sherman, Julia A., 122
Shklovsky, Victor, 144
Shlain, Leonard, 103, 123, 136, 138
Shneidman, Edwin S., 136
Silver, Ruth, 136
Silverman, Julian, 142
Simmel, Georg, 133
Simons, Daniel J., 4, 7, 46, 96, 97, 98, 99, 100
Singer, Peter, 123
Sloboda, John, 131, 138
Smith, Barbara H., 148
Smith, Edward E., 130
Sofsky, Wolfgang, 112
Sorensen, Roy, 104
Sowden, Paul T., 118, 132
Spelke, Elizabeth S., 104
Spender, Dale, 136
Spitzer, Gabriel, 131
Srinivasan, Narayanan, 137
Stalnaker, Robert, 111
Steinkraus, Warren E., 120
Stevens, Martin, 113, 117
Stockwell, Peter, 145
Stone, Christopher, 127

Stoppard, Tom, 87, 146
Storm, Ron W., 137
Subirana-Vilanova, J. Brian, 101, 105
Sullivan, Harry S., 100
Sykes, Gresham M., 100
Sypher, Wylie, 118

Taatgen, Niels A., 137
Tagg, Philip, 106
Taussig, Michael, 129
Teale, William H., 130, 142
Teshigahara, Wafu, 147
Thayer, Abbott H., 32, 33, 114, 115, 117
Thayer, Gerald, 32, 114, 115, 117
Thompson, Sandra A., 108
Titchener, Edward B., 11, 101, 102, 103
Tobin, Jacqueline L., 116
Townsend, James T., 137
Townsend, Jeanne, 97, 143
Trancik, Roger, 102, 103, 146
Treisman, Anne M., 137
Troscianko, Tom, 113, 114, 117, 118
Trubetzkoy, Nikolai S., 108
Truffaut, François, 120, 128
Tuan, Yi-Fu, 148
Tzu, Sun, 47, 120

Urry, John, 112, 123, 127, 129
Uskul, Ayse K., 124
Uspensky, Boris, 148

Van Peer, Willie, 144
Van Valkenburg, David, 103
Varela, Francisco J., 138, 139, 140
Varzi, Achille C., 102, 105, 109
Vaznis, Bill, 116

Vega-Bermudez, Francisco, 98, 99, 141
Vernon, Magdalen D., 11, 101, 102, 103, 105, 111

Wachtel, Paul L., 2, 97
Wackermann, Jiří, 111
Wallace, Alfred R., 32, 33, 114, 115
Wapner, Seymour, 104, 147
Wårvik, Brita, 104, 108, 109
Watson, Colin, 113
Watson, Ian, 123, 129
Watts, Alan, 101
Waugh, Linda, 108, 109
Weisstein, Naomi, 103
Welin, Stellan, 142
Wells, Herbert G., 115, 117
Werner, Heinz, 104, 147
Wever, Ernest G., 102
Williams, A. Mark, 131
Williams, Henry S., 115
Wiseman, Richard, 120
Witkin, Herman A., 50, 54, 118, 119, 121, 122, 124
Wittgenstein, Ludwig, 27, 112
Wolfe, Jeremy M., 117
Wong, Eva, 103
Wood, Beverly P., 132
Wood, Lindsay, 116

Zakia, Richard D., 101, 105, 109, 114, 121, 145, 146, 147
Zangemeister, W. H., 131
Zerubavel, Eviatar, 95, 100, 106, 107, 108, 109, 112, 119, 121, 122, 123, 127, 128, 129, 133, 136, 137, 143, 144, 149
Zerubavel, Noga, 100
Zerubavel, Yael, 105

SUBJECT INDEX

accompaniment, x, 18–19, 76
agenda setting, 62, 70
ambiguity, 86
analytical thinking, 51, 55
attention, 1–93
  allocation of, 3, 55
  capturing, 25, 46, 50, 57
  conceptual, 5, 7, 10, 20, 50, 75, 79, 92
  confined, 63, 68, 93
  deflecting, 45
  distributed, 75
  escaping, 28
  excessive, 81
  grabbing, 47
  paying, 2, 4, 25, 55, 57–59
  perceptual, 5, 7, 10, 20, 50, 92
  politics of, 8, 47–48, 59
  redirecting, 64
  scholarly, 28, 58–59, 62–63, 68
  shifting, 81, 83
  social organization of, 49–71
  switching, 5, 76–77
  See also attracting attention; collective attention; divided attention; focal attention; focused attention; joint attention; mnemonic attention; moral attention; multifocal attention; objects of attention; public attention; selective attention; simultaneous attention; underfocused attention; unfocused attention
attentional avoidance, 7, 60, 79
attentional battles, 57
attentional bias, 52–53, 63
attentional commonality, 51
attentional communities, 10, 52–53, 59, 63, 65
  membership in, 10, 52–55, 59, 63, 65, 67–68
attentional conventions, 10, 53, 56, 63–64, 90–91
  reversing, 8, 67, 81, 83
attentional deviance, 59, 61
attentional deviants, 60, 63–64, 81
attentional division of labor, 51–52
attentional ghettos, 63

SUBJECT INDEX

attentional habits, 8, 52–53, 63–64
   profession-specific, 56, 67–68
attentional marginalization, 28
attentional mentors, 64
attentional norms, 9–10, 53, 59–64, 81
attentional patterns, 8, 28, 51, 53–54, 66
   profession-specific, 66
attentional prominence, 11, 14, 19, 67, 71
attentional shelf life, 71
attentional socialization, 10, 63–69, 90
attentional span, 81
   collective, 71
attentional styles, 10, 50, 54–56, 82
   profession-specific, 56–57
attentional subcultures, 56
attentional taboos, 60, 62
attentional traditions, 10, 52–53, 56, 63–64, 68–69, 81
attentional variability, 51, 54, 56–57, 66
attention anchor, 2
attention deficit disorder (ADD), 81
attentiveness, 4, 56, 81
attracting attention, 14, 19, 25, 28, 32, 36, 53, 57, 76
autism, 51, 82
awareness, 3–4, 7, 20–21, 26–28, 50–51, 62, 66, 75, 78, 84–85, 89
   all-inclusive, 78
   margins of, 63, 77
   modes of, 2, 75, 78–79, 92
   over-inclusive, 73
   reduced, 7, 72
   restricted, 2, 7
   See also narrowed awareness; open awareness; peripheral awareness; public awareness; widened awareness

background, 7–93
   "in the," x, 5, 27, 29, 59, 82
background activities, 21
backgrounded features, 67
backgrounding, 45–46
backgroundless figures, 12
background-like character, 8, 14, 77, 79–81, 83, 86–87, 89–90
background-like surroundings, 7, 12, 17–19, 21, 24, 51, 54, 74, 77, 90–92
background matching. *See* camouflage
background music, 19–20
background persons, 28–30. *See also* extras
backing, 19
blending, 17, 30–32, 34, 37–38, 92
blinders, 52, 93
   intellectual, 63
blindness, 4–5, 56. *See also* color blindness; inattentional blindness; tactful blindness
blind spots, 6
   collective, 10, 71
borders, 86. *See also* boundaries; contours; edges; outline
boundaries, 15. *See also* borders; contours; edges; outline
boundedness, 15, 38
boundlessness, 15–16
brain, 9, 20, 52, 56, 72–73
bystanders, 28
"busy" design, 38, 40. *See also* perceptual complexity

camouflage, 6, 27–48
   art, 35
   background-matching, 30–41, 45–46
   contour-distorting, 37–45

SUBJECT INDEX | 195

dazzle, 39
  military, 30, 34–35, 38–39, 47
  See also diversion
censorship, 62
closed-mindedness, 74
clutter, 40, 74, 82
cognition, 2, 5–7, 9, 20, 24–25, 43, 50, 67–68
cognitive asymmetry, 11, 14, 22–23
cognitive control. *See* sociomental control
cognitive subversion, 8, 35–36
collective attention, 9–10, 69–71
coloration patterns, 31–34, 38
color blindness, 30–31
coloring books, 12, 90
concealing, 31–34, 36, 47. *See also* hiding
concentrating, 2, 4, 74–75, 77–78, 85
concern, 52, 64, 66, 81
  circle of, 52, 58
  focus of, 66
concerns, 50
  collective, 9
  curbing, 64
  moral, 5, 57, 64
  professional, 9, 51, 56, 66
cones, 77, 92
consciousness, 1–3, 72–73, 78
conspicuousness, 29–34
context, 13, 50, 55–57, 65, 75
contour distortion. *See* camouflage
contours, 8, 14, 16, 26, 30, 37–42, 44–45, 54, 76, 80, 86, 90
  blurring, 35
  perceptual dissolution of, 30. *See also* borders; boundaries; edges; outline
contrast, 26–27, 30–31, 38
convention, 8, 10, 12, 17–19, 21–22, 53, 56, 59–60, 63–65, 76, 79–83, 86–87, 89–92

creativity, 79
Cubism, 40, 42–43, 76
curiosity, 62, 66

decontextualization, 51, 56
defamiliarization, 84
delineation, 8, 14–15, 26, 52, 63, 76, 81, 86, 90
denial, 6, 75
detachability, 8, 16, 43, 51, 54–55, 74, 90–91
detecting, 26, 33, 40, 58, 74, 82
differentiating, 11, 17, 20, 24, 26, 31, 50–51, 54–55, 75, 82
disattention, 60–61, 63, 68–69, 75, 81
  norms of, 60
discreteness, 8, 50, 55, 91
discriminability, 26, 78, 92
disregarding, 9, 22, 24, 57, 60, 62, 68, 75
distinctness, 2, 4, 14, 17–18, 30, 34–35, 54, 92
distractibility, 79, 81
distraction, 2, 47, 52, 73–74, 78–79
diversion, 46–48
divided attention, 52, 75

edges, 29, 38, 86. *See also* borders; boundaries; contours; outline
Embedded Figures Test, 43, 50
embeddedness, 17, 22, 24, 26, 30–31, 40, 43, 46, 55, 65
empty space, 8, 12–13, 68, 85–87, 90
envisioning, 6, 8, 12, 15, 54, 63, 68, 90–91
epistemic revolutions, 58
essentialism, 8–9, 79, 81, 89–90, 92
eventfulness, 20–21
exclusion, 2–3, 60, 78

experts, 40, 64, 66–67
extraneousness, 8, 50, 81
extras, 28, 60
eye catching, 12–13, 25, 38
eye-openers, 84
eyes, 25–27, 31, 33, 47, 49, 52, 56, 60, 65, 73, 77, 79, 84, 92

familiarity, x, 14, 26–27, 43–44, 55, 58, 82, 84
field-dependence, 50–52, 54–55
field-independence, 50–52, 54–55
figure-and-ground model of perception, 7–9, 16–17, 20
figure-and-ground structure of the phenomenal world, 24
figure-ground reversal, 81, 83
figure-like character, 8, 12, 14–15, 17–21, 24, 31, 40, 43, 50–51, 54, 67, 74, 76–77, 79–83, 86, 89–91. *See also* objects of attention; thingness
figures, 7–93
filters, 6, 51, 53, 73, 79
finding, 25, 30, 64, 67, 74
focal attention, 11, 18, 67, 84
focused attention, 2, 4, 5, 7, 52, 54–55, 73, 75, 78, 92
focusing, 2, 4–5, 7–10, 16, 18, 20, 28, 46, 50–51, 53–57, 63, 68–69, 74–75, 78, 82, 84–85, 92
  mental, 8, 20, 54
focusing events, 71
focus of attention, 3–5, 68–70, 73–74
  collective, 70
  narrowing, 50, 77–78
foreground, 18, 21, 56, 66–67, 83
foregroundability, 81
foregrounding, 8, 82–85, 87, 89
framing, 21, 91–92

ganzfeld, 26
gaze, 2, 28, 51, 60, 62, 91
gestalt switch, 83
Gestalt theory of perception, ix, 7–8, 90
"gorilla experiment," 4, 7, 46
ground. *See* background

habituation, 8, 17, 27–28, 84, 90, 93
hearing, 1, 5, 17, 20, 37, 63, 72, 83
hidden figures, 43–45, 50, 54, 82, 84
hidden in plain sight, 1, 10, 44, 48, 93
hiding, 29, 31, 33, 36, 44, 46
hole-like character, 12, 33, 56, 90
holism, 55, 66
horizons, 78
  mental, 92
  moral, 58
hyperattentiveness, 53
hypervigilance, 81

ignorance, 6
  feigning, 62
ignoring, 3–4, 6, 8–11, 20–23, 27–28, 35, 50–53, 55–66, 68–69, 74–75, 78–79, 81–83, 85, 90–92
  collective, 9–10, 69–70
implicitness, 7, 22–23, 56, 59, 63–64, 68–69, 71, 74, 84
inattention, 4–5, 7, 10, 60, 68–70, 75, 81, 84
  auditory, 5
  civil, 60
  feigning, 61
  tactile, 5
inattentional blindness, 4–5
inattentiveness, 1, 53

in-betweenness, 14–15, 17, 64, 68, 85–87, 90
inconspicuousness, 29–32, 34
intermediate space (interspace), 12, 15, 90
invisibility, 6, 16–17, 28–29, 31, 35–37, 45, 84
  smell, 35
  social, 28–29, 71
irrelevance, 6–9, 21–22, 24, 28, 36, 50–53, 57, 62–64, 66–69, 72–75, 79, 81
  erotic, 52
  moral, 9, 52, 59, 64

joint attention, 9, 69

"key" information, 64, 81
knock-knock jokes, 44

language, 65, 67
listening, 18, 20, 26, 50–51, 62, 64, 67, 75–77, 83
logic, 8–9, 75, 81
looking, 1, 3, 5, 44, 46, 50, 58, 64, 72, 74, 78–79, 85
  refraining from, 60, 62, 78
looking for, 24–25, 40, 57, 67, 74
low profile, 29

"main idea," 62, 64, 81
markedness, 22–23, 29, 58, 74
mass media, 69–71
media coverage, 71. See also mass media
meditating, 2, 78
mental constriction, 74, 79
mental illumination, 3, 24, 78

mental readiness, 58, 79
mindfulness, 78
"minor" characters and details, 20, 62, 64, 71, 81, 87
mnemonic attention, 59
modernism, 87, 91–92
moral attention, 20, 52
  norms of, 59
multifocal attention, 75–77
multitasking, 75
"must-reads," 63
"must-sees," 62

namelessness, 22
naming, 22, 65, 84–85
narrowed awareness, 2, 49–51, 63, 72, 74, 77–78
narrow-mindedness, 74–75, 93
"negative" space, 12, 68, 85–87
noise, 24, 58
  background, 21, 24, 35–37, 46, 65, 67, 69
non-screeners, 50
non-thingness, 90
normality, 22, 29, 81
noteworthiness, 10, 22–23, 53, 57, 59, 64–65
noticeability, 6, 26–27, 29–31, 45, 58, 83
noticing, 1–2, 4–11, 14, 20–21, 23, 26–28, 31, 46, 49–52, 54–69, 72, 74, 79, 83–86, 91
novices, 66–67, 76

objects of attention, 2, 4, 7, 15, 19, 22, 26, 28–29, 31–32, 35–36, 38–40, 43, 46, 49–52, 54–58, 66–67, 74, 76–78, 82–83, 85–86, 90, 92. See also figure-like character; thingness

obsessive-compulsive disorder (OCD), 81
one-pointedness, 2
open awareness, 77–79, 92
openly covert communication, 36–37. See also steganography
optical puns, 86
ordinariness, 22, 27–29, 36, 57, 67, 84
outline, 38–40. See also borders; boundaries; contours; edges; silhouette
out-of-focus, 5
out-of-frame, 21, 91

perceptibility, 4, 14, 17, 31, 38, 53
perception, 1–2, 4, 11–12, 14–21, 26, 30, 38, 44, 46, 50, 55, 65, 67, 72, 74, 77, 79, 84, 92
  auditory, 7, 44, 76
  gustatory, 7
  olfactory, 7
  tactile, 7
  unconscious, 92
  visual, 7, 17, 35
perceptual complexity, 18, 38, 40, 43, 50, 54, 65, 67, 82
perceptual field, 11, 14, 18, 24, 26
perceptual homogeneity, 26, 40
perceptual readiness, 53
perceptual support, 11, 13, 19–21
peripheral awareness, 77, 85
peripheral hearing, 20
peripheral vision, 6, 67, 77, 85, 92–93
phenomenology, 11, 20, 24, 76–77, 92
polyphony, 26, 67, 76–77
predators, 29, 33–34, 40–41, 52
prey animals, 29, 31, 33–35, 40–41, 49

prominence, 19, 67, 71, 74
  attentional, 11, 14
  perceptual, 8
public agenda, 70
public attention, 57, 70–71
public attention cycle, 71
public awareness, 47, 71

radar screen
  collective, 70–71
  moral, 58
  public, 71
recognizability, 14, 16–17, 26, 38–39, 43, 72, 75, 86
refocusing, 83
relevance, 6–9, 21–22, 25, 50–51, 53, 57, 62–64, 67, 69, 73–74, 79, 81–82
  collective sense of, 9, 70
  moral, 52, 58
  professional sense of, 9, 56, 66
  public, 71
remarkability, 22–23, 28
rods, 77, 92
Rubin vase, 16, 18

salience, 25–26, 38, 49, 59, 70–71
  perceptual, 25, 74, 86
scanning, 24–26, 46, 66–67
schizophrenia, 73
scope of attention, 52, 81
  curbing others', 62–63
  widening, 78–79
scope of concern, 66, 81
screeners, 50
screening out, 6, 25, 50, 73, 79. See also tuning out
search image, 25
searching, 1, 24–25, 29, 31, 44, 74
searchlight, 24

SUBJECT INDEX | 199

seeing, 1, 14–15, 24–25, 29–31, 33, 38–39, 44, 49, 58, 64–65, 67–68, 72, 74–75, 78, 80, 83–85, 90
selective attention, 2–3, 5, 50, 53–54, 56, 70, 72–75, 77–79, 92
sense organs, 1, 3, 9, 49, 51–52, 72
senses, 2, 49, 73
separateness, 18, 38, 76
separating, 3, 8, 12, 17, 28, 31, 54, 63–64, 82, 90
shape, 14–16, 26, 30, 33, 38–39, 43, 80, 82, 85–86
   breaking up, 37, 39
shapelessness, 14–17, 80
signals, 24, 35–37, 46, 67
silhouette, 38, 86. *See also* contours; outline
simultaneous attention, 16, 18, 54, 56, 69, 75–76. *See also* multifocal attention
smell, 1, 17, 35, 63
socialization, 10, 63, 65, 90
   professional, 66–69
   See also attentional socialization
social problems, 57, 70, 84
socio-attentional patterns, 53
sociology of attention, 9, 53, 69
sociomental acts, 10, 63
sociomental control, 62
soundtrack, 20, 90
spotlight, 3–4, 83
spotlighting, 70, 83
spotting, 24–27, 29–31, 38, 40, 43–46, 50, 54–55, 67–68, 82
spying, 30
standing out, 2, 18, 25, 29–30
steganography, 35–37
streaming, 17–18
surroundings, 7–8, 12–13, 17, 24, 30–35, 37, 40, 46, 49–51, 54, 74–75, 77, 82, 84, 90–92

tacitness, 27, 59, 62–63, 65, 68–69, 81, 84
tact, 53, 60–61, 63
tactful blindness, 61
tactlessness, 61
taken-for-grantedness, 27, 84
targets, 24, 30–31, 34, 39–41, 46, 52, 76
taste, 1, 4, 17, 64
thingness, 12, 14, 26, 54, 74–75, 90–91. *See also* figure-like character, objects of attention
touch, 5, 7, 67
triage, 67
tuning out, 18, 28, 72
tunnel vision, 75, 93

unawareness, 1, 6, 36
underfocused attention, 81
unfocused attention, 81, 92
unmarkedness, 22–23, 28, 74, 84

vigilance, 6, 52, 66, 81
visibility, 1, 6, 17, 22, 26, 65
vision, 6, 9, 52, 68
   central, 77, 92
   field of, 29, 51–52
   mental, 93
   range of, 9, 51
   See also peripheral vision

"wagging the dog," 48
Waldo, 44, 50
widened awareness, 79, 93

zeroing in, 25, 52
zooming in, 74

Communal self.
alone - thoughts, memories, desires.